60 **Topics in Current Chemistry**
Fortschritte der chemischen Forschung

Structure of Liquids

Springer-Verlag
Berlin Heidelberg New York 1975

This series presents critical reviews of the present position and future trends in modern chemical research. It is addressed to all research and industrial chemists who wish to keep abreast of advances in their subject.

As a rule, contributions are specially commissioned. The editors and publishers will, however, always be pleased to receive suggestions and supplementary information. Papers are accepted for "Topics in Current Chemistry" in either German or English.

ISBN 3-540-07484-8   Springer-Verlag Berlin Heidelberg New York
ISBN 0-387-07484-8   Springer-Verlag New York Heidelberg Berlin

Library of Congress Cataloging in Publication Data. Main entry under title: Structure of liquids. (Topics in current chemistry; 60) Bibliographie: p. Includes index. CONTENTS: Schuster, P., Jakubetz, W., and Marius, W. Molecular models for the solvation of small ions and polar molecules. — Rice, S. A. Conjectures on the structure of amorphous solid and liquid water. 1. Solvation. 2. Molecular association. 3. Polywater. I. Schuster, Peter, 1941-Molecular models for the solvation of small ions and polar molecules. 1975. II. Rice, Stuart Alan, 1932- Conjectures on the structure of amorphous solid and liquid water. 1975.   III. Series. QD1.F58 vol. 60 [QD543]   540'.8s[541'.34]   75-29235

Contents

Molecular Models for the Solvation of Small Ions and Polar Molecules*

Prof. Dr. Peter Schuster, Dr. Werner Jakubetz, and Dr. Wolfgang Marius

Institut für Theoretische Chemie der Universität Wien, Währingerstraße 17, A-1090 Wien, Österreich

Contents

---

* Dedicated to Prof. H. Hartmann on the occasion of his 60th birthday

I. Introduction

In general, any ordered arrangement of solvent molecules surrounding a solute that differs significantly from the bulk structure of the pure liquid solvent is called a solvation shell. Depending on the forces acting between solute and solvent molecules, or between solvent molecules alone, the dynamic stability of a solvent shell can vary from a rather rigid structure, with its own normal modes of vibration and very low exchange rates for solvent molecules, to a flickering cluster where the molecules change place with almost the same frequency as in the bulk. As is well known from the phenomenon called hydrophobic interaction, a solvent shell with characteristic properties can also exist in cases where there are almost no specific interactions between solute and solvent molecules, but a highly ordered structure in the bulk of solvent. This kind of solvation has been most thoroughly investigated in water. Nevertheless, most experimental results concerning this subject lack any really satisfactory explanation, since hydrophobic interaction cannot be discussed reasonably on a molecular level without a detailed knowledge of the solvent's molecular structure, which is not known today for associated liquids like water.

Now, we might be attempted to turn our attention to the opposite extreme where the interaction between solute and solvent molecules overwhelms the interaction among solvent molecules alone. Solvation of small mon atomic ions represents an excellent example of this situation. The strong electric field of the ion provides an electrostatic basis for solute–solvent interaction. For a given solvent molecule, the electrostatic energy of interaction is largest in this example. It might be appropriate again to distinguish two different types of solvents. 1. The pure solvent does not show a high degree of order. Consequently, the directional properties of the ion dominate in a rather large region around the center and decrease gradually as we proceed into the unperturbed bulk. This situation is indicated in Fig. 1 A, where the solution is decomposed into an ordered zone — the solvation shell (a) — and the disordered bulk (b). 2. Solvents like

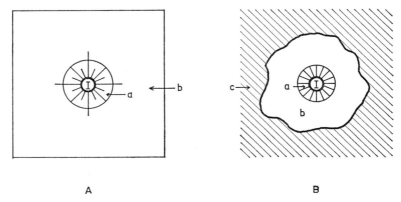

A   B

Fig. 1. Schematic models for ion solvation in simple solvents without ordered bulk (A) and in highly ordered solvents (B)

water with a highly ordered structure, however, behave very differently. A fairly simple model of ion solvation in water was proposed by Frank and Wen [1]. They distinguish between three different regions in the liquid surrounding the solute (Fig. 1 B). In the inner sphere (a) the solvent molecules are more or less strongly bound to the ion and therefore appear significantly less mobile than the molecules in the bulk. The structure of the pure solvent (c) will be encountered again at some distance from the ion. There is no reason to assume that the ordered structure of the solvation shell will merge into the structure of the bulk continuously without a break, in short range order. Therefore an intermediate region of disorder (b) with highly mobile water molecules has been introduced. Using this model, Frank and Wen [1] were able to explain the "structure making" and "structure breaking" properties of ions of different charge or size in aqueous solutions. Owing to the lack of molecular theories for associated liquids, any attempt at a detailed molecular description of region b or region c will at once encounter the problems mentioned above. The concept of different regions around the central ion has been generalized to some extent by Gurney[2] who used the expression "cosphere" for the region surrounding a spherical ion in which we have to expect significant differences in structure and properties of solvent molecules.

A further complication is introduced by the dynamics of the structures under consideration. A detailed knowledge of the most stable molecular structure of cospheres does not present an ultimate solution to the problem of ion solvation, since in many cases the lifetime of one particular configuration is not much longer than a few picoseconds [3]. The macroscopic properties, therefore, are not determined by one particular structure but by a complicated dynamic system. Consequently, a straightforward investigation of ionic solvation based on the Hamiltonian of the whole system leads to an intractable degree of complexity.

The solvation of polyatomic ions or polar neutral molecules is even more difficult to describe. There are two sources of additional problems: first of all, the symmetry of the system under investigation is drastically reduced and hence the number of different configurations increases tremendously. Furthermore, the strength of the electric field is much smaller than in the case of monatomic ions with spherical symmetry and therefore the dynamic behavior of the solvation shell is even more important for *a priori* calculations of macroscopic properties.

Many attempts have been made to calculate important macroscopic properties of ionic solutions without a detailed knowledge of their molecular structure. All models developed for this purpose can be classified into three categories according to the way the surrounding of the central ion is accounted for. Usually, continuous and discontinuous theories of ionic solvation are distinguished. Furthermore, it seems appropriate also to mention various attempts to apply statistical mechanics directly to ionic solutions.

1. The ions are regarded as rigid balls moving in a liquid bath. It is assumed that the macroscopic laws of motion in a viscous medium hold, and that the electrostatic interaction is determined by the theory of continuous dielectrics. This assumption implies that the moving particles are large compared to the molecular structure of the liquid. The most successful results of continuous theories can be found in any textbook of physical chemistry: Stokes', law for viscous motion, Einstein's derivation of the dependence of viscosity on the concentration

of the solute, the Born equation determining the free enthalpy of ion solvation, and Debye's investigation of rotational diffusion. Despite the fact that the assumption mentioned above is by no means fulfilled for small ions, these theories work very well provided the intrinsic parameters of the model have been carefully adjusted. But in continuous theories such "local" macroscopic quantities as local dielectric constants ecc. are introduced as parameters. From a more rigorous point of view these quantities are not well defined and look a little artificial.

2. In the approach usually called the discontinuous or chemical theory of solvation, the central ion and its nearest neighbors, constituting the so-called first solvation shell, are regarded as a rigid entity moving in the bulk liquid of the solvent. This kind of treatment, which dates back to Bernal and Fowler [4], proved successful in determining the thermodynamic properties of single ions in different solvents. The original model was modified and refined by Verwey [5]. A more recent calculation of free enthalpies of hydration and coordination numbers in the first hydration shell was reported hy Morf and Simon [6]. They applied Born's equation to the hydrated complex of the central ion and obtained very satisfactory results for a large variety of cations. Yet, the chemical model can only be a crude approximation, since there is no such strict distinction between first and higher solvation shells in the actual structure of solvated ions. It does, however, represent a step forward from continuous models toward a more rigorous treatment.

Statistical models for ionic solutions differ from an exact calculation based on quantum mechanics by the replacement of the full nonrelativistic Hamiltonian describing motions and interactions of electrons and nuclei by a model Hamiltonian, which represents the interactions between molecules as a whole. Initially, statistical mechanics was applied mainly to the interaction between dissolved particles, although the original cluster expansion of the grand partition function by Mc Millan and Mayer [7] was done in a very general form. For two more recent articles in this field see Friedman [8] or Jones and Mohling [9]. Additionally, the paper of Stecki [10] should also be mentioned here. He started from classical electrostatic and polarization energies of interaction between two particles and applied statistical mechanics to them. For some rigid-lattice models he was able to derive fairly complicated equations for the calculations of average forces acting on particles and for single ion free energies. In general the major problem in the statistical treatment is two-fold because neither distribution functions nor potentials for two or more particles are known or obtainable by simple experiments. An *a priori* determination of distribution functions of higher order involves very complicated calculations and therefore we might conclude that a purely statistical approach to ion solvation is not tractable nowadays for practical use even in cases where largely simplified two-body potentials are considered.

During the last few years the progress of computational techniques has made it possible to simulate the dynamic behavior of whole ensembles consisting of several hundred molecules. In this way the limitations of the statistical approach can be at least partly overcome. Two kinds of methods — molecular dynamics and Monte Carlo calculations — were applied to liquids and liquid mixtures and brought new insight into their structure and properties. Even some important characteristics of systems as complicated as associated liquids like water could be explained. Two recent papers, one for each type of calculation, are mentioned here [11,12]. For further details see also the references quoted therein. A more popular summary was presented recently by McDonald and Singer [13]. Hopefully we might expect that similar numerical calculations on the motion of particles around a central solute molecule or ion will bring some progress in the theory of solvation.

Recent developments in the theory of interaction between ions or polar molecules and their nearest neighbors in the solution are summarized in the present review. Furthermore, new experimental data are presented as long as they can be compared reasonably with calculated results for isolated complexes. After collecting the material available we found it appropriate to divide the discussion of data and results into two parts: ions and polar molecules. Most attention is given here to the ionic solvation since interactions between polar molecules, especially hydrogen bonding, have already been reviewed extensively [14-19].

The discussion on molecular solvation is restricted here to the structures and properties of a few particle associations which are important in solutions of ions and polar molecules. Certainly, we are aware of the fact that theory is still a long away from providing a satisfactory description of structures and energies of large clusters occurring in the liquid. Nevertheless, progress has been achieved in a few particle interactions during the last few years and further development seems promising. We feel that any relevant discussion of dynamic properties of molecular complexes has to start from sufficiently accurate energy surfaces which will probably be accessible in the near future.

Some other theoretical aspects of ionic solvation have been reviewed in the last few years. The interested reader is referred to them: ionic radii and enthalpies of hydration [20], a phenomenological approach to cation–solvent interactions mainly based on thermodynamic data [21], relationship between hydration energies and electrode potentials [22], dynamic structure of solvation shells [23]. Brief reviews, monographs, and surveys on this subject from a more or less different point of view have also been published [24-28].

An enormous amount of experimental material reflects the importance of solvation phenomena for understanding both chemical reactions in solution and the nature of the liquid state itself. Most of the accessible experimental techniques have been applied in solvation studies and occasionally different experimental approaches for the measurement of physical properties of solutions seem to have evolved as disciplines in their own right. Accordingly, the experimental aspects of solvation have also been reviewed from a great variety of points of view. A survey of the relevant literature shows that really comprehensive treatments of solvation can be found only in monographs devoted to solutions or solvents [25, 26, 28-31]. Review articles, which are more or less general and must necessarily be very concise, concentrate mostly on selected aspects of the whole subject, thus the publications of Rosseinsky [22], Hertz [23], Hinton and Amis [32, 33], Burgess and Symons [34], Blandamer and Fox [35], Blandamer [36], Grunwald and Ralph [37], Case [38], and Kecki [39] are mentioned here as representatives of a number of similar articles. Ionic solvation is also a preferred subject of periodic reports and review series [40-44] in which extensive bibliographies might be found.

P. Schuster, W. Jakubetz, and W. Marius

II. Theoretical Concepts of Molecular Interactions

The theory of molecular interactions has been reviewed extensively in the past and many monographs and contributed volumes are available in this field. Three of them[45-47] are especially recommended because they give the same kind of insight into the problems that is needed here. We do not intend to present an extensive discussion on the methods applied to intermolecular forces in this paper. The interested reader will be referred to some selected articles. Nevertheless it seems necessary at least to introduce the type of notation we prefer here and to stress some points that we feel to be important for a critical evaluation of the calculated results.

The various types of successful approaches can be classified into two groups: empirical model calculations based on molecular force fields and quantum mechanical approximations. In the first class of methods experimental data are used to evaluate the parameters which appear in the model. The shape of the potential surfaces in turn is described by expressions which were found to be appropriate by semiclassical[a] or quantum mechanical methods. Most calculations of this type are based upon the electrostatic model. Another more general approach, the "consistent force field" method, was recently applied to the forces in hydrogen-bonded crystals [48, 49].

The second class of theories can be characterized as attempts to find approximate solutions to the Schrödinger equation of the molecular complex as a whole. Two approaches became important in numerical calculations: perturbation theory (PT) and molecular orbital (MO) methods.

A. Electrostatic Models for Molecular Interactions

The electrostatic theory of interaction between ions or molecules was developed many years ago. The individual particles are assumed to be more or less rigid balls. The forces acting on them are determined by Coulomb's classical law of electrostatics. Summaries can be found in many textbooks, see *e.g.* [25, 45, 46].

More recently, electrostatic theory has been revived due to the concept of molecular electrostatic potentials. The potential of the solute molecule or ion was used successfully to discuss preferred orientations of solvent molecules or "solvation sites" [50-54]. Electrostatic potentials can be calculated without further difficulty provided the nuclear geometry (R_k) and the electron density function $\varrho(R)$ or the molecular wave function $\Psi(r_1\sigma_1, r_2\sigma_2 \ldots r_n\sigma_n)$ is known[b]:

$$\varrho(R_1) = n\int\Psi^+(r_1\sigma_1\ldots r_n\sigma_n)\, \Psi(r_1\sigma_1\ldots r_n\sigma_n)\, dr_2\ldots dr_n d\sigma_1\ldots d\sigma_n; \tag{1}$$

$$R_1 \equiv r_1$$

---

a) In this context the expression "semiclassical" is used for any method which is based on a quantum mechanical treatment of the electronic structure of individual molecules or ions and on calculations of intermolecular energies and forces by classical mechanics and electrostatics.

b) Unless stated otherwise atomic units are used in this article, see *e.g.* [55]. Therefore the mass of the electron, m_0, the elementary charge, e_0, and Planck's constant, $\hbar = h/2\pi$, equal 1. The two most important conversion factors for lengths and energies are: 1 a(tomic) l(ength) u(nit) $= 0.5292$ Å, 1 a(tomic) e(nergy) u(nit) $= 27.21$ eV $= 627.5$ kcal/mole.

By $\mathbf{r}_i = (x_i y_i z_i)$ and σ_i we denote spatial and spin coordinates of individual electrons. n represents the number of electrons in the molecule. For many cases it is appropriate to include nuclear charges as well and to define a molecular charge distribution $Q(\mathbf{R})$:

$$Q(\mathbf{R}) = -\varrho(\mathbf{R}) + \sum_{k=1}^{N} Z_k \delta(\mathbf{R} - \mathbf{R}_k) \tag{2}$$

By Z_k and $\mathbf{R}_k$ charges and positions of the N nuclei in the molecule are denoted. Now the electrostatic potential of molecule A can be formulated as follows (3):

$$V_A(\mathbf{R}) = \int \frac{Q(\mathbf{R}_1)}{|\mathbf{R} - \mathbf{R}_1|} d\mathbf{R}_1 = -\int \frac{\varrho(\mathbf{R}_1)}{|\mathbf{R} - \mathbf{R}_1|} d\mathbf{R}_1 + \sum_{k=1}^{N} \frac{Z_k}{|\mathbf{R} - \mathbf{R}_k|} \tag{3}$$

Starting from electrostatic potentials $V(\mathbf{R})$ the electrostatic or coulomb energy of interaction between two particles ΔE_{COU} is obtained by Eq. (4).

$$\Delta E_{\text{COU}}(\mathbf{R}) = \int Q_B(\mathbf{R}) V_A(\mathbf{R}) d\mathbf{R} = \int Q_A(\mathbf{R}) V_B(\mathbf{R}) d\mathbf{R} =$$

$$= \iint \varrho_A(\mathbf{R}_1) \varrho_B(\mathbf{R}_2) \frac{1}{|\mathbf{R}_1 - \mathbf{R}_2|} d\mathbf{R}_1 d\mathbf{R}_2 + \sum_{k \in A} \sum_{l \in B} \frac{Z_k Z_l}{|\mathbf{R}_k - \mathbf{R}_l|} \tag{4}$$

$$- \sum_{k \in A} \int \varrho_B(\mathbf{R}_1) \frac{Z_k}{|\mathbf{R}_1 - \mathbf{R}_k|} d\mathbf{R}_1 - \sum_{l \in B} \int \varrho_A(\mathbf{R}_1) \frac{Z_l}{|\mathbf{R}_1 - \mathbf{R}_l|} d\mathbf{R}_1$$

Here $\mathbf{R}$ represents the configurational freedom of the molecular complex. In the case of two rigid particles there are six degrees of freedom. Consequently, even for the interaction between two particles $\Delta E_{\text{COU}}(\mathbf{R})$ is a highly complicated energy surface. Symmetry arguments of course may reduce the number of different configurations, e.g. for two linear molecules $\mathbf{R}$ consists of only four degrees of freedom.

Since a very recent review article on the use of electrostatic potentials became available in this series [19], it is unnecessary to go into details here. Nevertheless, one comment seems important to us and should be added here. As we will see in the following subsections, electrostatic theory neglects various important contributions to the total energy of interaction $\Delta E(\mathbf{R})$. Relative energies and equilibrium geometries of many important complexes, however, are predicted correctly by electrostatic calculations. With regard to energies of interaction the main reason for this success was certainly brought about by compensation of contributions with a different sign (see p. 25). In most of the examples investigated $\Delta E_{\text{COU}}(\mathbf{R})$ comes close to $\Delta E(\mathbf{R})$ around the energy minima. Another important result from electrostatic theory concerns intermolecular orientations. In many associations of polar molecules or ions the preferred geometry is determined by the angular dependence of electrostatic forces [50] and thus correct equilibrium geometries are obtained. There are, of course, many exceptions where the electrostatic predictions fail [17,18,56].

In general, electrostatic theory is not able to reproduce correct energy surfaces as a whole since the compensation of errors mentioned above holds only around the energy minima (see p. 25). Furthermore, the various spectral data, especially IR, RAMAN and NMR results, which became accessible during the last ten years cannot be interpreted at all by electrostatic models. Many theoreticians feel, therefore, that there is a need for more accurate calculations on ion–molecule complexes based upon quantum mechanics.

B. Quantum Mechanical Calculations on Intermolecular Interactions

During the last decade MO-theory became by far the most well developed quantum mechanical method for numerical calculations on molecules. Small molecules, mainly diatomics, or highly symmetric structures were treated most accurately. Now applicability and limitations of the independent particle, or Hartree-Fock (H. F.), approximation in calculations of molecular properties are well understood. An impressive number of molecular calculations including electron correlation is available today. Around the equilibrium geometries of molecules, electron-pair theories were found to be the most economical for actual calculations of correlation effects[c]. Unfortunately, accurate calculations as mentioned above are beyond the present computational possibilities for larger molecular structures. Therefore approximations have to be introduced in the investigation of problems of chemical interest. Consequently the reliability of calculated results has to be checked carefully for every kind of application. Three types of approximations are of interest in connection with this article.

The first kind of simplification exclusively concerns the size of the basis set used in the linear combination of one center orbitals. Variational principle is still fulfilled by this type of *"ab initio"* SCF calculation, but the number of functions applied is not as large as necessary to come close to the H. F. limit of the total energy. Most calculations of medium-sized structures consisting for example of some hydrogens and a few second row atoms, are characterized by this deficiency. Although these calculations belong to the class of *"ab initio"* investigations of molecular structure, basis set effects were shown to be important [54] and unfortunately the number of artificial results due to a limited basis is not too small.

In semiempirical calculations variational principle itself is manipulated, or to give another interpretation, the molecular Hamiltonian is modified in order to save computation time. Two semiempirical methods differing in the concept and the degree of sophistication have been applied frequently to ion–molecule complexes and therefore will be mentioned briefly.

The CNDO procedure proposed and developed by Pople and his co-workers [59–61], as well as a number of its variants, are characterized by the assumption of vanishing overlap between basis set orbitals (CNDO = Complete Neglect of Differential Overlap), which reduces the number of non-zero two electron integrals drastically. Various other approximations and modifications had to be introduced

---

[c] The interested reader is referred to two recent surveys of accurate calculations on molecules [57] and electron-pair theories [58]. A very brief introduction can also be found in Ref.[18].

in order to retain rotational invariance of the calculated results and to guarantee reasonable interatomic distances in molecules. Recent reviews of CNDO and related methods can be found in two monographs [62,63], to which the interested reader is referred.

The second kind of semiempirical procedure mentioned here is even cruder. In the extended Hueckel method (EHM [64,65]) the electronic structure of the molecule is simulated by an effective Hamiltonian. The total energy of the molecule is represented by a sum of one electron energies and even the nuclear repulsion terms are not taken into account explicitly. This type of approach can be shown to give an approximate idea of electronic structures and relative energies of unpolar molecules like hydrocarbons, but it fails inevitably when applied to structures with appreciable polarity [66]. Therefore any application of EHM calculations to interactions between polar molecules or ions should be regarded with a good deal of scepticism.

Quantum mechanical calculations of intermolecular forces generally start from wave functions of the isolated particles. With regard to the actual treatment of the interaction, however, there is some competition between perturbation theory and MO methods.

1. Perturbation Theory

Although intermolecular perturbation theory was formulated many years ago [67], it is still being developed further to a higher and higher degree of perfection, see e.g. [68–72]. For our purpose a less general and less sophisticated approach derived some years ago by Murrell and his co-workers [73–76] will be sufficient. This kind of perturbational treatment of intermolecular forces was applied in actual numerical calculations on various systems: $(He)_2$ [77,78], $(He)_3$ [79], and models for hydrogen bonding ([80] and references therein). In this article we will use perturbation theory more because of its heuristic merits in explaining the nature of different contributions to the total energy of interaction ΔE [d].

In intermolecular perturbation theory one of the major problems concerns electron exchange between molecules. In the method described here exchange is limited to single electrons. This simplification is definitely a good approximation at large intermolecular distances. The energy of interaction between the molecules, $\Delta E(R)$, is obtained as a sum of first order, second order, and higher order contributions:

$$\Delta E = \underbrace{\Delta E_{\text{COU}} + \Delta E_{\text{EX}}^{(1)}}_{1^{\text{st}} \text{ order terms}} + \underbrace{\Delta E_{\text{POL}} + \Delta E_{\text{DIS}} + \Delta E_{\text{CHT}}}_{2^{\text{nd}} \text{ order terms}} \cdots \tag{5}$$

---

d) In this paper ΔE is always defined as the energy of complex formation:

$$A + B \rightarrow AB; \ \Delta E = E_{AB} - (E_A^0 + E_B^0)$$

E_A^0, E_B^0 are total energies of the isolated molecules A and B; E_{AB} represents the total energy of the associated, or supermolecule, AB. E_{AB} and ΔE, as mentioned before, are functions of various conformational degrees of freedom: $E_{AB} = E_{AB}(R)$, $\Delta E = \Delta E(R)$.

The less important second order contributions and higher order contributions are not shown explicitly in Eq. (5). The superscript of exchange energy indicates that only single–electron exchange has been taken into account.

The first order perturbation energy is calculated by the following zeroth-order approximation to the ground-state wave function Ψ^0 of the two molecules:

$$\Psi^0 = N\,(1+P)\,\Psi^0_A\,\Psi^0_B.\tag{6}$$

N represents a normalization constant of the wave function. P takes care of antisymmetric behavior with respect to electron exchange. In the approximation of single-electron exchange, P can be substituted by

$$P \simeq \sum_{i\in A}\sum_{j\in B} P_{ij}\tag{7}$$

where P_{ij} represents a permutation operator interchanging the coordinates of the electrons i and j. i and j are assumed to belong originally to the molecules A and B, respectively. The wave functions of the isolated molecules are denoted by Ψ^0_A and Ψ^0_B. They are solutions to the corresponding Schrödinger equations of the Hamiltonians $\mathscr{H}^0_A$ and $\mathscr{H}^0_B$:

$$\mathscr{H}^0_A\Psi^0_A = E^0_A\Psi^0_A;\ \mathscr{H}^0_B\Psi^0_B = E^0_B\Psi^0_B\,.\tag{8}$$

The total Hamiltonian $\mathscr{H}$ of the supermolecule AB is split into $\mathscr{H}_0$ and the perturbation $\lambda\mathscr{H}'$. In all actual calculations on molecular interactions the intensity parameter λ is set $\lambda=1$:

$$\mathscr{H} = \mathscr{H}_0 + \lambda\mathscr{H}';\ \ \mathscr{H}_0 = \mathscr{H}^0_A + \mathscr{H}^0_B;\ \ \mathscr{H}' = V_{AB}(R);\tag{9}$$

V_{AB} represents the operator for the intermolecular potential energy (10):

$$V_{AB}(R) = -\sum_{i\in A}\sum_{l\in B}\frac{Z_l}{|r_i - R_l|} - \sum_{j\in B}\sum_{k\in A}\frac{Z_k}{|r_j - R_k|}$$
$$+ \sum_{i\in A}\sum_{j\in B}\frac{1}{|r_i - r_j|} + \sum_{k\in A}\sum_{l\in B}\frac{Z_k \cdot Z_l}{|R_k - R_l|}\,.\tag{10}$$

Using Eqs. (6) and (10) the first-order contribution to intermolecular perturbation energy can be derived without further difficulty [73,74] :

$$\Delta E_{\mathrm{COU}} + \Delta E^{(1)}_{\mathrm{EX}} = \frac{U_{00}+U'_{00}}{1+S'_{00}} \simeq U_{00} + U'_{00} - U_{00}S'_{00} - U'_{00}S'_{00}\,.\tag{11}$$

The integrals are defined in the following way:

$$U_{00} = \int (\Psi^0_A\,\Psi^0_B)^+\,V_{AB}(R)\,\Psi^0_A\,\Psi^0_B\,d\tau\tag{12}$$

$$U_{00}' = \int (\Psi_A^0 \, \Psi_B^0)^+ \, V_{AB}(R) P(\Psi_A^0 \, \Psi_A^0) d\tau \tag{13}$$

$$S_{00}' = \int (\Psi_A^0 \, \Psi_B^0)^+ \, P(\Psi_A^0 \, \Psi_B^0) d\tau \, . \tag{14}$$

S_{00}' is exactly and U_{00}' is roughly proportional to the square of overlap between individual molecular orbitals of both subsystems A and B. The dependence of U_{00}' on orbital overlap has been discussed in detail [75]. The last term in Eq. (11), $U_{00}' S_{00}'$, is roughly proportional to the fourth power of overlap between orbitals and hence much smaller in magnitude at great enough intermolecular distances. Therefore it is usually added to the higher order contributions.

More or less arbitrarily, the electrostatic energy of interaction ΔE_{COU} is defined directly in the isolated molecules. The renormalization correction of the electrostatic energy $(U_{00} S_{00}')$ is included in the exchange term $\Delta E_{EX}^{(1)}$.

$$\Delta E_{COU} = U_{00}; \; \Delta E_{EX}^{(1)} = U_{00}' - U_{00} S_{00}' \tag{15}$$

An alternative partitioning, *e.g.* $\Delta E_{COU} = U_{00}/(1 + S_{00}')$ and $\Delta E_{EX}^{(1)} = U_{00}'/(1 + S_{00}')$ could also be justified because there are no physical criteria for the preference of one or the other assignment [e]. The only advantage of the choice in Eq. (15) is that terms of same order of overlap between individual molecular orbitals are collected.

Furthermore, in this scheme of partitioning the formula for ΔE_{COU} becomes identical with the semiclassical electrostatic energy of interaction (4). As we shall see later a closely related quantity can be defined in the framework of intermolecular MO theory and thereby comparison of different approaches is largely facilitated.

For two closed shell molecules, A and B, the first-order exchange contribution always shows a positive sign and represents the non-classical repulsion between molecules or ions. Actually it is roughly proportional to the square of the overlap between molecules or orbitals and hence falls off exponentially with increasing distance. Simple classical electrostatic models usually assume rigid spherical particles. Repulsion between rigid balls is represented by an infinite step function for potential energy. We may now identify this simple discontinuous repulsive potential as a very crude approximation to the exponential decrease of exchange energy.

The three major second-order contributions to perturbation energy can be interpreted as changes in the energy of interaction due to first-order changes in the wave functions. Usually, changes in the wave functions are described by partial excitations to the unoccupied or virtual orbitals [f] of the isolated molecules. Three different cases can be distinguished (Fig. 2).

---

[e] The difficulty here arises from the nonvanishing overlaps between the molecules. At sufficiently large intermolecular distances overlap and exchange energy become negligibly small and then the electrostatic energy of course is a physically well-defined quantity.

[f] Here and in the following considerations we refer more to actual calculations than to the rigorous formalism of perturbation theory. For practical reasons the molecular excited states are simulated by raising electrons into the higher, virtual energy levels obtained by the MO—SCF procedure.

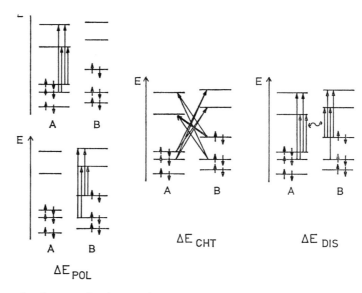

Fig. 2. Second-order contributions to intermolecular perturbation energies (schematic description by orbital excitations within the framework of the independent particle model; ΔE_{POL} and ΔE_{CHT} are represented by single excitations, ΔE_{DIS} by correlated double excitations)

1. Polarization of molecule A in the field of molecule B, or vice versa. These two contributions are summed up as the polarization energy ΔE_{POL}.

2. Charge transfer from one molecule to the other is described by partial excitations of an electron from an occupied orbital of one molecule to a virtual orbital of the other. The energetic contribution of this intermolecular delocalization of electrons is called charge transfer energy, ΔE_{CHT}. In the theory of complex ions charge transfer energies are usually interpreted as covalent contributions to the metal ligand bond. As we shall see, straightforward definitions of charge transfer are very hard to find.

3. Simultaneous and correlated excitations of electrons in both molecules lead to correlation of electron motions and to a general net stabilization of the complex. This effect is usually attributed to the so-called attractive dispersion forces and the corresponding energy is therefore called dispersion energy ΔE_{DIS}.

In general, all three contributions to the second order perturbation theory stabilize the molecular association AB. ΔE_{POL} is the analogue to the classical polarization energy and becomes identical with the function $(\sum_{n} C_n \cdot |R|^{-n})$ calculated from multipole moments and molecular polarizabilities at large intermolecular distances ($|R|$). ΔE_{CHT} on the other hand is roughly proportional to the square of overlap between molecular orbitals and hence falls off much more rapidly with increasing $|R|$ than the other two contributions. The dispersion energy ΔE_{DIS}, of course is identical with the London or attractive Van Der Waals energy. Again the separation of the second-order energy into individual contributions, especially

12

into charge transfer and polarization energies is only free of arbitrariness at long distances where molecular overlap is sufficiently small. It is easy to visualize that in strongly overlapping systems any reasonable distinction between charge transfer and polarization disappears, because in this case all molecular orbitals are more or less extended over the whole complex. We have to keep in mind, however, that at intermediate distances partitioning of perturbation energies into individual contributions always depends to some extent on the model assumption applied.

The overlap between molecular orbitals of different particles in the complex introduces a serious problem in the theory of molecular interactions. Furthermore, overlap between molecules or ions is a good criterion for the presence of non-classical charge transfer or covalent interactions in the complex. In fact, large intermolecular overlap or a high amount of charge transfer makes it impossible to recognize individual subsystems without arbitrariness. For a rough orientation we therefore calculated overlap integrals between valence orbitals of those atoms in the subsystems, A and B, which come closest together at the equilibrium geometry of the complex. These overlap integrals give an idea of the extent of intermolecular overlap. The numerical values obtained from Slater-type atomic orbitals (STO) are shown in Table 1. For elements with more than one valence orbital we calculated overlap integrals from s and p_σ functions as well as from their three most important hybrids: sp, sp^2, and sp^3. Although the numbers obtained differ significantly, general conclusions do not depend on the particular shape of the valence orbital chosen.

Not unexpectedly, different types of molecular associations differ largely with respect to intermolecular overlap. In fact overlap is very small in associations of rare gas atoms and unpolar molecules. On the other hand, molecules forming hydrogen bonds overlap rather strongly in the complexes. Consequently, there is a striking difference between the $F^-\ldots HOH$ and the other ion–water complexes considered here. In $F^-\ldots HOH$ a strong hydrogen bond is formed. Additionally, due to the stronger electrostatic forces stabilizing the complex and thereby reducing the intermolecular distance, intermolecular overlap is even larger than in the water dimer. Nevertheless, charge transfer effects are expected to have more relative importance in the latter case, since the total energy of interaction in $(H_2O)_2$ is much smaller than in $F^-\ldots HOH$. Among anion-molecule complexes $F^-\ldots HOH$ is exceptional, since no strong hydrogen bond is formed in the other anion–water complexes. As Table 1 shows, intermolecular overlap in $Cl^-\ldots HOH$ is significantly smaller than in $F^-\ldots HOH$. Overlap between the subsystems in cation–molecule associations like $Li^+\ldots OH_2$, $Na^+\ldots OH_2$, and $K^+\ldots OH_2$ is indeed very small. In $Be^{2+}\ldots OH_2$ a somewhat larger value is obtained. Interestingly, even in ionic molecules like LiF a significantly smaller value for the overlap integral than in hydrogen-bonded complexes was found. We can therefore expect covalent contributions to be of minor importance in all these types of complexes.

Before we turn to MO theory of molecular interactions a short discussion on the reliability of semiempirical calculations of the CNDO type by means of perturbation theory would be useful. For a better understanding of the possibilities and limitations of semiempirical MO approaches to intermolecular forces we calculated first–order perturbation energies for very simple complexes with and

Table 1. Overlap between valence orbitals at the two closest atoms A and B in some equilibrium geometries of molecular complexes

Structure	Atoms A	B	R_{AB} (Å)	R_{AB} (a.u.)	ΔE (kcal/mole)	S_{AB}[1] S_B	$P_{\sigma B}$	h_{1B}	h_{2B}	h_{3B}
He...He	He	He	2.96	5.59	− 0.022	0.0030				
He...HF	He	H	2.25	4.26	− 0.151	0.0665				
H_2O...HOH	H	O	2.02	3.82	− 3.90	0.1097	0.0961	0.1455	0.1418	0.1381
F^-...HOH	H	F	1.57	2.96	−23.7	0.2045	0.1624	0.2594	0.2506	0.2429
Cl^-...HOH	H	Cl	2.97	5.61	−11.9	0.0554	0.0660	0.0858	0.0859	0.0849
H_2O...Be^{2+}	Be	O	1.56	2.95	−141.4	0.0244	0.0400	0.0455	0.0467	0.0468
H_2O...Li^+	Li	O	1.89	3.57	−35.2	0.0165	0.0260	0.0301	0.0308	0.0308
F^-...Li^+	Li	F	1.51	2.85	−150.0	0.0455	0.0686	0.0807	0.0823	0.0821
						$S_A\,S_B$	$P_{\sigma A}\,P_{\sigma B}$	$S_A\,P_{\sigma B}$	$P_{\sigma A}\,S_B$	$P_{\pi A}\,P_{\pi B}$
H_2O...Na^+	Na	O	2.25	4.25	− 24.0	0.0061	0.0115	0.0100	0.0071	0.0015
H_2O...K^+	K	O	2.69	5.08	− 16.6	0.0089	0.0187	0.0127	0.0129	0.0023

[1]) Overlap integrals were calculated from Slater-type orbitals: $\phi_A = \phi_{1S}$ at H, He, Li^+, or Be^{2+}, $\phi_A = \phi_{2S}$, $\phi_{2P\sigma}$ for Na^+, and $\phi_A = \phi_{3S}$, $\phi_{3P\sigma}$ for K^+; $\phi_B = \phi_{1S}$ at H or He, $\phi_B = \phi_{2S}$, $\phi_{2P\sigma}$ at O or F^- and $\phi_B = \phi_{3S}$, $\phi_{3P\sigma}$ at Cl^-. Hybrid orbitals are defined as follows:

$h_{1B} = \frac{1}{\sqrt{2}} (S_B + P_{\sigma B})$, $h_{2B} = \frac{1}{\sqrt{3}} (S_B + \sqrt{2}\,P_{\sigma B})$, and $h_{3B} = \frac{1}{2} (S_B + \sqrt{3}\,P_{\sigma B})$. Orbital exponents: $\xi_H = 1.0$, $\xi_{He} = 1.7$, $\xi_{Li^+} = 2.7$, $\xi_{Be^{2+}} = 3.7$, $\xi_O = 2.275$, $\xi_{F^-} = 2.425$, $\xi_{Na^+} = 3.425$, $\xi_{Cl^-} = 1.916$, and $\xi_{K^+} = 2.583$.

without the approximations of the CNDO/2 procedure. Further details of these calculations are given elsewhere [18, 81].

At a first glance we are confronted with a striking contradiction: Differential overlap is neglected in the CNDO method due to the ZDO approximation (16). Therefore there is no exchange contribution to the energy of interaction in CNDO or related calculations, $\Delta E_{\mathrm{EX}}^{(1)} = 0$.

$$\phi_a^+ \phi_b d\tau = \delta_{ab} |\phi_a|^2 \, d\tau$$

$$S_{ab} = \int \phi_a^+ \phi_b d\tau = \delta_{ab} \tag{16}$$

$$(ab/cd) = \int\int \phi_a^+(1) \, \phi_c^+(2) \, \frac{1}{|\boldsymbol{r}_1 - \boldsymbol{r}_2|} \, \phi_b(1) \phi_d(2) d\tau_1 d\tau_2 = \delta_{ac} \delta_{bd}(aa/cc)$$

The exchange contribution in an *ab initio* perturbation theory is the only repulsive term[g] around the energy minimum in most of the stable complexes and consequently we would expect no net repulsion between two closed shell molecules in semiempirical calculations. On the other hand it is known from actual calculations that intermolecular interactions are described more or less correctly by the CNDO/2 procedure. Indeed, strong repulsion is obtained between closed shell molecules. Evidently there must be another approximation which compensates for the neglect of exchange energy. With regard to the simplifications of the CNDO/2 method we find that this is in fact the case. The approximation shown in Eq. (17) is responsible for the repulsive term.

$$\int \phi_a^+(1) \, U_{\mathrm{B}} \phi_a(1) \, d\tau_1 = -Z_{\mathrm{B}} \int \phi_a^+(1) \, \frac{1}{|\boldsymbol{r}_1 - \boldsymbol{R}_{\mathrm{B}}|} \, \phi_a(1) \, d\tau_1 =$$

$$= -Z_{\mathrm{B}}(aa/\mathrm{R}_{\mathrm{B}}^{-1}) \simeq -Z_{\mathrm{B}}(aa/bb) \tag{17}$$

The two electron coulomb integral (aa/bb) is smaller than the point charge integral $(aa/\mathrm{R}_{\mathrm{B}}^{-1})$ at all finite distances and therefore the use of Eq. (17) introduces an underestimation of an attractive contribution to the intermolecular energy of interaction or, if we put it the other way round, a net repulsion between the molecules results.

In any case it seems worthwhile to consider a particular example. For the sake of simplicity we have chosen a small complex, with only four electrons, which can be described approximately by Slater-1s orbitals. Since we are interested mainly in ion–molecule interactions, the system Li$^+$...He seems appropriate for our purpose[h]. From Eqs. (11)—(15) the following expressions can be derived without further difficulty:

$$\Delta E_{\mathrm{COU}} = 2 \{2(aa/bb) - Z_{\mathrm{B}}(aa/\mathrm{R}_{\mathrm{B}}^{-1}) - Z_{\mathrm{A}}(bb/\mathrm{R}_{\mathrm{A}}^{-1})\} + \frac{Z_{\mathrm{A}} \cdot Z_{\mathrm{B}}}{R} \tag{18}$$

$$R = |\boldsymbol{R}_{\mathrm{A}} - \boldsymbol{R}_{\mathrm{B}}|$$

---

[g] Repulsion due to the nuclear interaction energy becomes important only at much shorter interatomic distances.

[h] The interaction between two He atoms was analyzed in the same manner [18]. Very similar results were obtained.

$$\Delta E_{\mathrm{EX}}^{(1)} = -2\{(ab/ab) + S_{ab}[(aa/ab) + (ab/bb) - Z_{\mathrm{B}}(ab/R_{\mathrm{B}}^{-1}) - Z_{\mathrm{A}}(ab/R_{\mathrm{A}}^{-1})] - \\ - S_{ab}^{2}[3(aa/bb) - Z_{\mathrm{B}}(aa/R_{\mathrm{B}}^{-1}) - Z_{\mathrm{A}}(bb/R_{\mathrm{A}}^{-1})]\} \tag{19}$$

Introducing the approximations of the CNDO/2 procedure into the equations for the first-order energies we obtain:

$$\Delta E_{\mathrm{COU}}^{\mathrm{CNDO/2}} = [4 - 2(Z_{\mathrm{A}} + Z_{\mathrm{B}})] (aa/bb) + \frac{Z_{\mathrm{A}} \cdot Z_{\mathrm{B}}}{R} \tag{20}$$

$$\Delta E_{\mathrm{EX}}^{\mathrm{CNDO}} = 0 \tag{21}$$

The numerical values at a number of distances, R_{AB}, were calculated for all the four expressions given above, ξ_a and ξ_b were represented by $1s$ orbitals centered at the nuclei A and B. The orbital exponents $\xi_a = 2.7$ and $\xi_b = 1.7$ were chosen according to Slater's rules for $1s$ orbitals at Li^+ and He, respectively. The dependence of the individual contributions on R_{AB} is shown in Fig. 3. The two errors

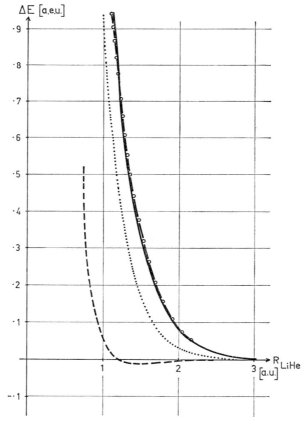

Fig. 3. Energy partitioning in (LiHe)+ obtained by first-order perturbation theory. (Basis set: $1s$(He), $\xi = 1.7$ and $1s$(Li+), $\xi = 2.7$; exact calculations: ΔE:———, ΔE_{COU}: –––––, and $\Delta E_{\mathrm{EX}}^{(1)}$: –o–o–o–. CNDO approximation: $\Delta E^{\mathrm{CNDO/2}} = \Delta E_{\mathrm{COU}}^{\mathrm{CNDO/2}}$:......)

introduced by the approximations of the CNDO/2 procedure partly compensate each other. In comparison to the He_2 system [18], however, a systematic underestimate of the repulsive contribution is found.

A corollary of our discussion on the use of semiempirical methods in the calculation of the intermolecular energies can be formulated as follows. CNDO/2 and related calculations might be very useful for the investigation of some types of intermolecular associations. Any agreement with experimental data, however, is brought about by error compensation. Energy partitioning in the framework of perturbation theory based upon semiempirical methods, on the other hand, has no meaning at all because the final results are determined exclusively by approximations and model assumptions of the method itself.

Finally, we have to beat in mind that the simple application of perturbation theory to intermolecular forces used here is valid only in the range of vanishing intermolecular overlap. At smaller intermolecular distances double and multiple exchange of electrons between the interacting particles becomes important [78]. Furthermore, partitioning of the energy of interaction becomes artificial, e.g. the stabilizing contribution of ΔE_{COU} in the $Li^+ \ldots He$ system certainly has no physical meaning. It is simply a result of our convention of renormalization, Eqs. (11) and (15), and of course falls off to zero in the region of vanishing intermolecular overlap.

2. Molecular Orbital Theory

In principle direct MO calculations of intermolecular energies are free of almost all the errors mentioned in the preceding subsection. The energy of interaction is obtained as the difference between the total energies of the association and the isolated molecules:

$$\Delta E = E_{AB} - (E_A^0 + E_B^0) \tag{22}$$

The energy of the complex, AB, is calculated as an approximative solution of the Schrödinger equation in exactly the same way

$$\mathscr{H}\Psi_{AB} = E_{AB}\Psi_{AB} \tag{23}$$

as for isolated molecules. The major problem, however, arises from the numerical details. The energy of interaction is obtained as a difference between huge numbers. For example, in case of the water dimer ΔE amounts to only 0.005 % of the total energy. Hence all kinds of errors, methodical or numerical, might appear largely amplified in intermolecular energies. No appreciable progress in this field was possible before high accuracy and high speed computers became available. Nowadays MO theory can be applied to molecular interactions at various stages of sophistication. We will mention here three different levels: semiempirical all valence electron calculations, *ab initio* SCF calculations, and *ab initio* calculations including electron correlation.

An impressive number of semiempirical and *ab initio* SCF calculations on molecular associations of different kinds is available now. In many cases very accurate *ab initio* SCF calculations using extended basis sets were also performed.

On the other hand, accurate calculations beyond the level of the independent particle approximation are very rare. Recently, a detailed study of $(HF)_2$ was published [82] in which the influence of electron correlation on the properties of the hydrogen bond was investigated.

In order to learn more about the nature of the intermolecular forces we will start with partitioning of the total molecular energy, ΔE, into individual contri butions, which are as close as possible to those we defined in intermolecular perturbation theory. Attempts to split ΔE into suitable parts were undertaken independently by several groups [83-85]. The most detailed scheme of energy partitioning within the framework of MO theory was proposed by Morokuma [85] and his definitions are discussed here[1]. This analysis starts from antisymmetrized wave functions of the isolated molecules, Ψ_A^0 and Ψ_B^0, as well as from the complete Hamiltonian $\mathscr{H}$ of the interacting complex AB. Four different approximative wave functions are used to describe the whole system:

1. A simple Hartree product of the two wave functions:

$$\Psi_1 = \Psi_A^0 \cdot \Psi_B^0 \tag{24}$$

2. An antisymmetrized Hartree-Fock product of the two wave functions

$$\Psi_2 = \mathscr{A}\,(\Psi_A^0 \cdot \Psi_B^0); \; \mathscr{A} = N \cdot (1+P) \tag{25}$$

[see Eq. (7)].

The operator $\mathscr{A}$ takes care of the antisymmetric property of the wave function and is usually called antisymmetrizer.

3. A product of the two wave functions, which have been optimized individ-ually in the potential field of the other molecule:

$$\Psi_3 = \Psi_A \cdot \Psi_B \tag{26}$$

Again Ψ_A and Ψ_B are antisymmetric.

4. The ordinary, antisymmetric and optimized SCF wave function of the complex

$$\Psi_4 = \Psi_{AB}. \tag{27}$$

Now expectation values of the energies are calculated from all the four wave functions. The individual contributions to the energy of interaction are obtained as differences:

$$\Delta E_i = E_i - (E_A^0 + E_B^0); \; E_i = \int \Psi_i^* \mathscr{H} \Psi_i d\tau; \; 1 \leqslant i \leqslant 4 \tag{28}$$

---

[1] Another type of energy partitioning in MO theory was proposed by Clementi [86]. This kind of partitioning called "Bond Energy Analysis" (BEA) was found to be particularly useful in larger molecular clusters [87-90]. Since the individual contributions of BEA are not directly related to the quantities discussed in intermolecular perturbation theory, Clementi's technique will not be used here (see Chapter V).

On examining the definitions given above more closely we find that ΔE_1 is equivalent to the electrostatic energy of interaction, ΔE_{COU}, defined in intermolecular perturbation theory, cf. Eq. (15), and becomes identical with it when the exact solution of the Schrödinger equations for the isolated molecules are used. ΔE_2 on the other hand is the analogue to the complete first-order energy of interaction including the exchange contribution. Accordingly, we can define an intermolecular exchange energy within the framework of MO theory:

$$\Delta E_{\text{EX}} = \Delta E_2 - \Delta E_1 \tag{29}$$

This exchange energy, however, differs somewhat from the usual exchange contribution in perturbation theory, since in the MO treatment the exchange is not restricted to single electrons. The one electron MO's in the wave function Ψ_3 account for the field of the other molecules whereas no electron exchange is allowed. Physically we might interpret this contribution as a sum of electrostatic and polarization energies[j]:

$$\Delta E_{\text{POL}} + \Delta E_{\text{COU}} = \Delta E_3; \quad \Delta E_{\text{POL}} = \Delta E_3 - \Delta E_1 \tag{30}$$

Finally the remaining difference to the total SCF energy of interaction, ΔE_4, is attributed to a stabilization of the complex as a consequence of electron delocalization or charge transfer between the molecules:

$$\Delta E_{\text{CHT}} = \Delta E_4 - (\Delta E_{\text{COU}} + \Delta E_{\text{EX}} + \Delta E_{\text{POL}}) = \Delta E_4 + \Delta E_1 - \Delta E_2 - \Delta E_3 \tag{31}$$

Actually, Morokuma [85] applied his procedure of energy partitioning to calculations on hydrogen-bonded complexes using Slater-Type orbitals (STO's). Since we will refer afterwards mainly to calculations using GTO basis sets, we tried to reproduce Morokuma's results with a GTO basis [91], simulating the outer region of STO's as closely as possible (STO-4G basis [92]). The results summarized in Table 2 show that intermolecular energies and their partitioning into individual contributions are obtained almost identical in the STO and the STO-4G basis, although the total energy of the isolated water molecule differs appreciably in both basis sets. Furthermore, the electrostatic contribution, ΔE_{COU}, comes fairly close to the total energy of interaction, ΔE.

In order to see whether this kind of intermolecular energy partitioning also works well in ion–molecule complexes, and to study basis set effects on the results, we performed a number of calculations on $\text{Li}^+\ldots\text{OH}_2$ [91]. The numerical values obtained are shown in Table 3. To facilitate comparison the geometry of the water molecule was frozen at the experimental values and the intermolecular distance was fixed to $R_{\text{OLi}} = 3.4$ a. l. u. $\simeq 1.8$ Å, which is very near to the equilibrium value.

---

[j] Morokuma [85] defined $(\Delta E_3 - \Delta E_1)$ as the sum of polarization and dispersion energy. If Ψ_A and Ψ_B are single Slater determinants as is usual in SCF calculations, no correlation effects are taken into account and hence this contribution actually represents electron polarization alone.

Table 2. Intermolecular energy partitioning in $(H_2O)_2$ at the equilibrium geometry obtained by an STO [85] and an STO—4G [91] calculation

Quantity[2]	STO-calculation	STO—4G[1] calculation
E_{H_2O} (a. e. u.)	−75.70503	−75.50009
ΔE_{COU}	− 8.00	− 7.82
ΔE_{EX}	9.86	9.70
ΔE_{POL}	− 0.25	− 0.26
ΔE_{CHT}	− 8.16	− 7.94
ΔE_{DEL}	− 8.40	− 8.20
ΔE	− 6.55	− 6.31

[1] The STO—4G basis set is described in Ref. [92].
[2] Energy differences in (kcal/mole).

Table 3. Intermolecular energy partitioning in $Li^+...OH_2$ at an intermolecular distance $R_{OLi} = 3.4$ a. l. u. ~ 1.8 Å obtained with different basis sets[1]

Quantity[2]	GTO-basis sets[1]				
	A (7,3; 3) Li^+: (6)	B (9,5; 4) Li^+: (7)	C (9,5; 4) Li^+ (7,1)	D (9,5,1; 4) Li^+: (7,1)	E (11,7,1; 6,1) Li^+: (7,1)
E_{H_2O}(a.e.u.)	−75.87760	−76.01012	−76.01012	−76.03711	−76.05308
E_{Li^+}(a.e.u.)	− 7.23306	− 7.23605	− 7.23605	− 7.23605	− 7.23605
ΔE_{COU}	−46.5	−52.0	−52.0	−44.2	−41.7
ΔE_{EX}	8.4	15.1	15.1	14.9	15.3
ΔE_{POL}	− 6.8	− 9.8	− 9.8	−11.3	−12.8
ΔE_{CHT}	− 1.0	3.8	3.3	3.4	5.2
ΔE_{DEL}	− 7.8	− 6.1	− 6.5	− 7.9	− 7.6
ΔE	−45.9	−43.0	−43.4	−37.3	−33.9
u_{H_2O}(D)	2.39	2.64	2.64	2.33	2.20

[1] The following short notation for basis sets will be applied throughout in this contribution: $(\alpha, \beta, \gamma/\lambda, \mu)$ indicates that α s-type, β p-type, and γ d-type orbitals are applied for second-row atoms and λ s-type and μ p-type function for H-atoms. Contractions are given in square brackets: $(\alpha', \beta', \gamma'/\lambda', \mu')$. Basis sets C, D, and E are almost identical to or exactly the same as the ones used in Ref. [109–111] or [86] respectively.
[2] Energy differences in (kcal/mole).

A comparison of energy partitioning in Tables 2 and 3 is really illustrative in explaining the main difference between a hydrogen-bonded association and a cation–molecule complex. In the first example ΔE_{COU}, ΔE_{EX}, and ΔE_{DEL} are of the same order of magnitude. Since the exchange term is always repulsive and therefore shows the opposite sign in comparison to the other two contributions, the resulting total energy of a hydrogen bond is obtained as a complicated super-position of various contributions, just as was suggested many years ago by Coulson and some other authors, e.g. [93,94]. As expected, in the ionic complex the energy of interaction is dominated by the electrostatic term (ΔE_{COU}). Exchange repulsion is very similar in both kinds of molecular interactions. According to the stronger electrostatic field of the cation polarization energy (ΔE_{POL}) is much larger in the ionic complex. Charge transfer, on the other hand, seems to be of major importance in the hydrogen-bonded association. The meaning of ΔE_{CHT} in MO energy partitioning is not completely clear if the numbers shown in Table 3 are compared. Depending on the basis set applied ΔE_{CHT} in $Li^+\ldots OH_2$ has a positive or negative sign. We shall return to this problem below when basis set effects will be discussed more extensively.

The results of energy partitioning in $Li^+\ldots OH_2$ obtained with a number of different basis sets are listed in Table 3. Since intermolecular overlap is small in these kind of complexes (Table 1), we expect the electrostatic model to be a good approximation for classical contributions to the total energy of interaction. Indeed, ΔE_{COU} is to a good approximation proportional to the dipole moment of the water molecule calculated with the same basis set. This can be seen even more clearly in Table 4 where ΔE_{COU} is compared with ion–dipole and ion–quadrupole energies obtained from the classical expression of the multipole expansion series [45,95-97]:

$$\Delta E_{COU}^{CL} = - \frac{q_A \cdot \mu_B}{R^2} + \frac{q_A \cdot \Theta_{zz}^B}{R^3} + \ldots \qquad (32)^{k)}$$

Table 4. Electrostatic contribution to intermolecular energies in $Li^+\ldots OH_2$ and $Li^+\ldots OCH_2$ at $R_{OLi} = 3.4$ a.l.u. ~ 1.8 Å (all energies in kcal/mole)

	Basis[1] set	ΔE_{COU}	$-\mu_{calc}/R_{CM}^2$	μ_{exp}/R_{CM}^2	$\Theta_{zz}^{exp}/R_{CM}^3$	ΔE_{COU}^{Cl}
$Li^+\ldots OH_2$	A	−46.5	−47.5			
	B	−52.0	−52.6			
	C	−52.0	−52.6	−36.8	−1.4	−38.2
	D	−44.2	−46.3			
	E	−41.7	−43.7			
$Li^+\ldots OCH_2$	C	−39.8	−36.2	−27.9	−3.7	−31.6
	D	−34.4	−33.4			

[1] The basis sets applied are defined in Table 3.

The intermolecular distance was calculated here relative to the center of mass (C. M.) of the water molecule. For this particular choice of the origin the quadrupole term is rather small and the total electrostatic energy of interaction is reproduced fairly well by the first term of the expansion, the ion–dipole term alone. The difference between ΔE_{COU} and ΔE_{COU}^{CL} of course is not exclusively

---

k) The expression for the classical electrostatic energy given here is derived for the interaction of molecule B with C_{2v}-symmetry, e.g. H_2O or H_2CO, with a point charge (q_A) situated on the C_{2v}-symmetry axis (z-axis). Θ represents the diagonal, traceless quadrupole tensor of B ($\Theta_{xx} + \Theta_{yy} + \Theta_{zz} = 0$).

due to neglect of higher terms in the multipole expansion. As shown previously, renormalization of the wave function as a result of intermolecular overlap introduces arbitrariness into the definition of the electrostatic term. This contribution to ΔE_{COU} is certainly small in our case and one can regard the type of MO energy partitioning applied here as a successful approach as far as the electrostatic part is concerned.

On the other hand, the values for ΔE_{COU} obtained with different basis sets reflect the relative weakness of SCF calculations in reproducing the electrostatic molecular potential. We note an appreciable improvement of μ and ΔE_{COU} when polarization functions are added to the basis set[1]. Comparing the calculated electrostatic energy with the value obtained from the experimental dipole- and quadrupole moments of water, we realize that the *ab initio* energy is larger even when a very extended basis set (cf. [87]) leading to a near H. F. energy was applied. This difference is simply a result of the fact that dipole moments of many polar molecules, like H_2O, H_2CO, CO etc., are reproduced rather poorly at the H. F. limit (cf. Tables 3 and 5 or [87]). Finally, it seems worthwhile to mention that an "intramolecular" effect on correlation energy of interaction [18,82] of the order of about 5 kcal/mole must be expected, which should result from the correlation correction to the molecular dipole moment of water.

Exchange repulsion does not seem to be as sensitive as the electrostatic contribution provided that the basis sets used are large enough. The addition of polarization functions to the basis set has no significant influence on ΔE_{EX}. On the other hand, very small basis sets, such as the smallest basis sets used here, cause ΔE_{EX} to be underestimated remarkably.

Polarization energy, ΔE_{POL}, shows a monotonic increase in absolute value on an extension of the basis set (Table 3). In particular, polarization functions at the atoms of the water molecule are very effective. The reliability of calculated polarizabilities at the H. F. limit does not seem to be completely clear. For hydrogen fluoride the calculated polarizability with a near H. F. wave function was found to be substantially too small [82]. The consideration of electron correlation leads to an increase in the calculated value which is in good agreement with experimental data. For the water molecule, however, Liebmann and Moskowitz [104] obtained a result very close to the experimental value using a variational procedure based upon a near H. F. wave function [100] (Table 5). Nevertheless, it seems interesting to compare the numbers shown in Table 3 with the value for classical polarization energy calculated from expression (33) derived for the polarization of a molecule B with C_{2v}-symmetry with a point charge q_A placed on the z-axis [97].

α_{zz}^{B} represents here the

$$\Delta E_{\text{POL}}^{\text{CL}} = -\frac{1}{2} \frac{q_A^2 \cdot \alpha_{zz}^B}{R^4} + \dots \tag{33}$$

---

[1] Polarization functions are best defined as atomic orbitals which are not occupied in the ground states of atoms due to symmetry, *e.g.* p, d orbitals at H or d, f orbitals at O atoms. They provide an additional flexibility of the basis set in the molecule, where the symmetry is significantly lower than in the atom.

Table 5. Properties of the isolated ligands H_2O and H_2CO[1]

Molecule	$\lvert R_{CM}-R_0 \rvert^2$ (Å)	Dipole moment $\mu(D)$	Ref.	Quadrupole moment[2] $(10^{-26}$ esu. cm$^2)$ Θ_{yy}	Θ_{zz}	Ref.	Polarizabilities[2] $(10^{-24}$ cm$^3)$ α_{xx}	α_{yy}	α_{zz}	$\bar{a}$	Ref.
H_2O	0.066	1.85 ± 0.03	98)	2.63 (2.53)	−0.13 (−0.11)	99) 100)	(1.226)	(1.651)	(1.452)	1.45 (1.443)	103) 104)
H_2CO	0.604	2.33 ± 0.02	98)	-0.6 ± 0.7 (0.25)	-0.75 ± 0.8 (0.006)	104) 102)	(2.313)	(2.710)	(4.553)	(3.292)	104)

1) Experimental data are given as far as available. Calculated results are shown in parentheses.
2) The molecular coordinate system was chosen such that the z-axis coincides with the C_{2v} symmetry axis of the molecule. The molecular plane is identical with the yz-plane of the coordinate system. CM represents the center of mass, which for symmetry reasons lies on the z-axis. $\lvert R_{CM}-R_0 \rvert$ is the distance between the center of mass and the oxygen atom of the ligand.

z-component of the polarizability tensor of B. Using the values shown in Table 5 we obtain a classical polarization energy of $\Delta E_{POL}^{CL} = -18.7$ kcal/mole [m)] for Li[+]...OH$_2$, the intermolecular distance being defined relative to the C. M. of H$_2$O, which agrees reasonably with the values from MO energy partitioning, provided a correction of $\alpha_{zz}^{H_2O}$ due to electron correlation is taken into account. The second term in the expansion series (33), resulting from hyperpolarizabilities, is expected to be smaller by at least one order of magnitude.

Finally, we are left with the most tricky problem of providing an explanation for basis set effects on ΔE_{CHT} as shown in Table 3. Starting from a negative value obtained with a rather poor basis set (A), the charge transfer term changes sign and becomes larger with increasing flexibility of the basis. Going back to our original definitions of various contributions we realize there is a significant difference between ΔE_{CHT} in perturbation theory and MO energy partitioning, respectively. In the former case, the charge transfer term was clearly defined by certain configurations added to the ground-state wave function. In the latter case, however, ΔE_{CHT} is more or less a collection of the remnants when the total energy of the SCF-calculation is partitioned into individually defined contributions, see Eq. (31). In addition to the second order exchange or charge transfer term other contributions, repulsive in nature, might be hidden in ΔE_{CHT}. Consequently, we feel that the question of charge transfer between the individual components of a molecular complex should be discussed independently of its energetic manifestation, by an analysis of the molecular wave function. We will therefore return to the concept of charge transfer at the end of the chapter. In any case ΔE_{CHT} is responsible for one interesting basis set effect which is very important indeed in the intermolecular MO theory. The basis set of a molecule in a complex is improved to some extent by the presence of the basis functions at the other constituents. Characteristically we might call basis sets without nuclei "ghost" molecules. These "ghost" molecules lead to an artificial net stabilization of the complex. For a detailed discussion of "ghost" effects in hydrogen-bonded association the interested reader is referred to [18)], where more references to papers on this subject are given. Recently, "ghost" basis effects in ion–molecule complexes were also calculated directly [106)]. The values for ΔE_{CHT} in Table 3 indicate that this artificial stabilization of the complex becomes less important on an extension of the basis set. Further discussions on basis set effects on calculated energies of interaction can be found in Chapter V.

The last important contribution to intermolecular energies that will be mentioned here, the dispersion energy (ΔE_{DIS}), is not accessible in H. F. calculations. In our simplified picture of second-order effects in the perturbation theory (Fig. 2), ΔE_{DIS} was obtained by correlated double excitations in both subsystems A and B, for which a variational wave function consisting of a single Slater determinant cannot account. An empirical estimate of the dispersion energy in Li[+]...OH$_2$ based upon the well-known London formula (see e.g. [107)]) gave a

---

m) Using classical formulas Spears [105)] calculated the following values for classical electrostatic and polarization energies in Li[+]...OH$_2$, which are very close to our results: $\Delta E_{COU}^{CL} = -36.6$ kcal/mole (dipole term: -34.7 kcal/mole, quadrupole term: -1.9 kcal/mole), and $\Delta E_{POL}^{CL} = -17.9$ kcal/mole.

value of $\Delta E_{DIS} = -0.4$ kcal/mole [105], thus indicating that dispersion forces are not extremely important in these kind of cation–molecule complexes. Only one calculation of correlation effects in $Li^+ \ldots OH_2$ has been reported until now [108]. A value of $\Delta E_{COR} = 1.0$ kcal/mole was obtained. Since no analysis of intra- and intermolecular correlation effects is available, an estimate of ΔE_{DIS} from this calculation is impossible.

In Fig. 4 MO energy partitioning in $Li^+ \ldots OH_2$ is shown as a function of intermolecular distance [109]. As expected, the exchange energy ΔE_{EX} falls off much faster than the two other contributions ΔE_{COU} and ΔE_{DEL}. (ΔE_{DEL} is used here for the sum of delocalization effects, $\Delta E_{POL} + \Delta E_{CHT}$.) Interestingly, the curves for the total energy of interaction and the electrostatic contribution cross at a point near the energy minimum, thus explaining the success of simple electrostatic models.

To give a second example the interaction of Li^+ with formaldehyde, a somewhat larger ligand containing highly polarizable π-electrons, is shown in Fig. 5 [109,110]. Comparison of Figs. 4 and 5 shows that the potential curve for $\Delta E(R_{OLi})$ is very

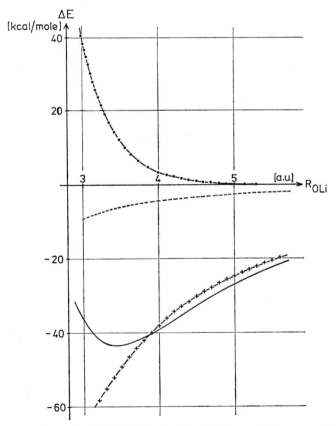

Fig. 4. Energy partitioning in $(LiOH_2)^+$ obtained by MO theory. (ΔE: ———— ΔE_{COU}: –+–+– ΔE_{EX}: –o–o–o– and ΔE_{DEL}: -----)

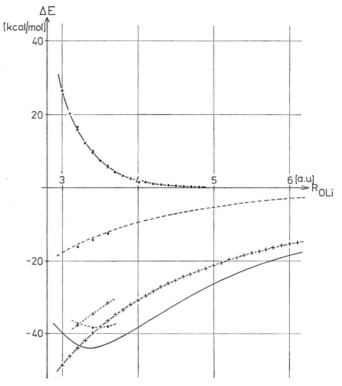

Fig. 5. Energy partitioning in $(LiOCH_2)^+$ obtained by MO theory (basis set C, ΔE: ——— ΔE_{COU}: −+−+− ΔE_{EX}: −·−·− and ΔE_{DEL}: −−−−; basis set D, ΔE: ·······, ΔE_{COU}: −−+−−+−−+, $\Delta E_{EX}, \Delta E_{DEL}$: ·)

similar for both systems, although the electrostatic contribution is substantially smaller in $Li^+ \ldots OCH_2$. At first sight this seems to be a striking contradiction to electrostatic considerations, since the dipole moment of formaldehyde is appreciably larger than the one of the water molecule (Table 5). But a more careful estimate accounting for quadrupole terms reveals that the apparent similarity is only brought about by the particular choice of the coordinate system. Since the quadrupole moment of H_2CO is fairly small only in a C.M. system (Table 5), we should really use the C.M. as the origin for our estimate. The classical formula (32) reproduces the MO results very well (Table 4). At the same distance from the O atom $(R_{OLi} = 3.4$ a.l.u. $\simeq 1.8$ Å) the Li^+ cation is much closer to the origin of the C.M. coordinate system in $Li^+ \ldots OH_2$ than in $Li^+ \ldots OCH_2$. Again we find two important results concerning the choice of basis sets: polarization functions (d orbitals at O and C) improve the dipole moment of the isolated H_2CO molecule significantly and have therefore a remarkable influence on the electrostatic contribution, ΔE_{COU}. Even very extended basis sets near the H.F. limit [102] cannot, however, reproduce the dipole moment of H_2CO correctly and consequently lead to an absolute value of ΔE_{COU} which is too large.

The dependence of ΔE_{COU} on the intermolecular distance, on the other hand, is reproduced very well in both cases by the simple electrostatic formula provided the calculated dipole moment was used for the ligand and the intermolecular distance was defined relative to its center of mass (Fig. 6).

The decrease of exchange energy with R_{OLi} is almost identical in both complexes discussed here (Figs. 4 and 5). A very careful estimate of the slopes shows the O atom in H_2CO to be a little less rigid than the one in H_2O.

From the presence of π-electrons in H_2CO we expect a higher polarizability, which is actually obtained by calculation [103]. Furthermore the polarization energy should be higher in $Li^+\ldots OCH_2$ than in $Li^+\ldots OH_2$. Comparison of the corresponding curves for ΔE_{DEL} in Figs. 4 and 5 shows indeed that polarization is much more important in the first example. An estimate of the polarization energy at the energy minimum ($R_{OLi} = 3.4$ a.l.u.) by the classical formula (33) gives a somewhat higher value than in case of $Li^+\ldots OH_2$ as well: $\Delta E_{POL}^{CL} = -22.8$ kcal/mole. Since no experimental value is available the polarizability of H_2CO was taken from the calculation of Liebmann and Moskowitz [104].

In the last section of this chapter we shall return to the problem of charge transfer in molecular complexes. We have just seen above that charge transfer energy as defined by MO energy partitioning (31) does not represent a good measure for the pure, energetic effect of charge transfer, since it also contains other contributions which are not directly related to electron delocalization

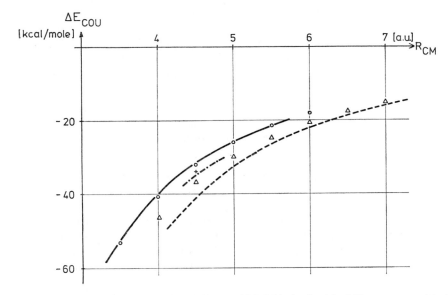

Fig. 6. Comparison of ΔE_{COU} in $(LiOH_2)^+$ and $(LiOCH_2)^+$ obtained by MO energy partitioning and classical electrostatics ($\Delta E_{COU}(LiOH_2)^+$: ——— (basis set C); $\Delta E_{COU}(LiOCH_2)^+$: ――――― (basis set C), and –·–·–·– (basis set D); ΔE_{COU}^{CL} is approximated by the first term in Eq. (32); $\mu_{H_2O}^{calc} = 2.64$ D (o; basis set C); $\mu_{H_2O}^{calc} = 3.02$ D (Δ, basis set C) and 2.79 D (+, basis set D); R_{CM} represents the distance between the Li nucleus and the center of mass of H_2O or H_2CO respectively)

between the molecules. Now we will make an attempt to analyze the wave function of the complex. No observable quantity is related directly to the amount of charge transferred. Therefore any definition of charge transfer inevitably implies some arbitrariness. Mulliken population analysis [112] has been frequently used to discuss charge transfer. Net atomic charges are summed up to give net charges of the molecules in the complex. The common procedure has been criticized frequently since the values obtained depend appreciably on the basis sets applied. An alternative analysis starts out from density matrices of the complex and the isolated subsystems [113]. Electron density maps provide a good idea of shape and size of a complex, but they are not useful for a discussion of chemical bonds or intermolecular forces. Details of electronic rerrangements are described more appropriately by density difference functions, which are obtained simply by substracting the densities of the isolated constituents from the density of the complex, the nuclear coordinates remaining unchanged (34).

$$\Delta \varrho(x, y, z) = \varrho_{AB}(x, y, z) - \{\varrho_A^0(x, y, z) + \varrho_B^0(x, y, z)\} \tag{34}$$

Typical examples of density- and density difference maps for ion–molecule complexes are shown in Chapter V. For our purpose we do not need the detailed spatial information which is rather confusing for a discussion of charge transfer. We proposed therefore [113] to choose an axis of the complex (z-axis), which is characteristic for the type of interaction that we are interested in. In $Li^+ \ldots OH_2$ or $Li^+ \ldots OCH_2$ this is evidently the twofold symmetry axis C_2. In other cases like $e.g.$ $F^- \ldots HOH$ or $(H_2O)_2$, the connection line of the three atoms forming the hydrogen bond will be appropriate[n]. After defining a z-axis in this way we can calculate a density difference curve $\Delta \varrho(z)$ by simple integration (35):

$$\Delta \varrho(z) = \iint \Delta \varrho(x, y, z) dx dy \tag{35}$$

In contrast to the three dimensional map, $\Delta \varrho(z)$ is much easier to analyze. Some examples of curves for different complexes are shown in Chapter V.

In order to calculate the charge transfer $\Delta \varrho$ (36) from one

$$\Delta q = \int_{-\infty}^{z_0} \Delta \varrho(z) dz = -\int_{z_0}^{+\infty} \Delta \varrho(z) dz \tag{36}$$

subsystem to the other we have to define a borderline between them. According to our type of analysis, a plane $z = z_0$ separating A from B in three dimensional space has to be chosen. Certainly this is not possible without some arbitrariness. In case of $(H_2O)_2$, however, we have shown [18], that different reasonable choices

---

[n] There are of course other types of complexes lacking global or local symmetry, where it is difficult to find a unique direction, $e.g.$ $Cl^- \ldots H_2O$ at its equilibrium geometry. In this case a multidimensional analysis of the density difference map might be the only reasonable way to define charge transfer.

of z_0 lead to almost the same results, especially if relative values are desired. Here we will use two different definitions for z_0:

1. In almost all examples of molecular complexes discussed here $\Delta\varrho(z)$ becomes zero somewhere in between both subsystems A and B (see Figs. 12–16, 18, 23). In our first definition we draw a borderline at this value of z:

$$\Delta\varrho(z_0) = 0$$

2. The other possibility to define z_0 starts from an integrated density function $\varrho(z)$, which is shown in Eq. (37). Now we define the border between A and B

$$\varrho(z) = \iint \varrho_{AB}(x, y, z)dxdy \tag{37}$$

at that point, where $\varrho(z)$ passes a minimum:

$$\varrho(z_0) = \text{min.}; \left(\frac{d\varrho(z)}{dz}\right)_{z_0} = 0 .$$

Choosing a plane as a border surface separating the subsystems is only a first approximation. One could go a step further and use the definitions given here analogously in three dimensions. This would lead to significantly different results, however, only in the case of larger complexes with complicated spatial structure, where a plane would be no longer a reasonable approximation to the border surface between the subsystems.

Calculated values for Δq in a number of molecular complexes are compared in Table 6. First of all we see that the amount of charge transferred is rather small in all the examples studied here. Numbers obtained from population analysis are significantly larger than those obtained from density functions. This difference is especially large in cation–water complexes where it reaches one order of magnitude. Both definitions of z_0, on the other hand, give very similar results. The amount of charge transferred depends on both the overlap between the subsystems and the charges of the interacting particles. As expected, Δq is much smaller in

Table 6. Charge transfer in molecular complexes obtained by different methods

Complex		Charge transfer $\Delta q(e_0)$		
Electron donor	Electron acceptor	Integration of density differences		Population analysis[1]
		$z_0 : \Delta\varrho_z = 0$	$z_0 : \varrho_z = \text{min.}$	
F^-	HOH	0.0190	0.0378	0.0542
H_2O	HOH	0.0032	0.0079	0.0110
H_2O	Li^+	0.0027	0.0040	0.0328
H_2O	Be^{2+}	0.0340	0.0740	0.3290

[1] Atomic populations and net charges defined according to Mulliken [112].

$Li^+..OH_2$ than in $F^-...HOH$, thus indicating the formation of a strong hydrogen bond in the latter case. The largest charge transfer is found in $Be^{2+}..OH_2$. Interestingly Δq in water dimer is not much larger than in $Li^+...OH_2$.

Finally, we would like to make a short comment on semiempirical calculations of the CNDO type. An analysis of CNDO/2 electron densities shows that among all complexes investigated a reasonable description is obtained only in hydrogen bonds between neutral molecules. In cation–water-[113] and anion–water complexes, the integrated density difference curves obtained by CNDO calculations deviate strongly from the *ab initio* curves in the domain of the ion (see also Chapter V), and the amount of charge transferred is grossly overemphasized. Values for Δq were obtained that were up to two orders of magnitude larger than those in Table 6. Population analysis gives the same qualitative results. The CNDO/2 wave function therefore is not a good approximation in these cases. In general we can conclude that the CNDO procedure is not able to describe ion–molecule complexes correctly, at least as far as electron distributions are concerned.

III. Experimental Background

We do not intend to add another review of solvation experiments to the great number already in existence. Such a project would be far outside the scope of this paper in which we give only some background information on solute–solvent interactions. In the absence of any satisfactory theory of the molecular structure of associated liquids, it is hard to see how the outcomes of purely molecular models could reasonably be compared, on a rigorous basis, with experimental results, which are of course influenced by interference from the bulk of the liquid. This is especially true for aqueous solutions, which are considered in most theoretical and also experimental contributions to the solvation problem. To close the gap left by the missing theory of associated liquids, it will usually be necessary to make use of phenomenological and macroscopic quantities or to introduce various parameters (which would obviously be related to macroscopic properties) in order to connect the predictions of molecular considerations with measured solution properties. Disregarding these demands, the huge amount of experimental material which has accumulated could act as a pitfall for theoreticians. Though it is obvious that for instance experimental solvation enthalpies cannot be compared in a straightforward manner with energies of interaction as obtained by molecular calculations, there are occasionally misunderstandings in the setup of the proper thermodynamic cycles required to relate two properties as different as the ones mentioned. Another source of ambiguity concerns calculations of experimentally only vaguely defined magnitudes like solvation numbers [33]. On the other hand, the existence of sometimes broadly spread spectra of experimental values also acts towards a philosophy of "good agreement with experiment", which might be either without any real significance or else does not actually increase our knowledge.

By the way, since single-ion solvation energies, as used for the characterization of solvation properties [22,36], cannot be determined directly from experiments, they must be derived through the adoption of additional conventions. Generally, Born-cycles [38,114] underlie the determination of such single ion properties and therefore proper ionic radii have to be established [20,115]. While it might seem that the demand for ionic radii offers a challenge for molecular calculations, this need not necessarily be the case. There is no reason why the definitions of ionic size required for the use in Born-cycles, where charged hard sphere-particles are assumed to interact with uniform dielectric fluids [114], should be identical with any of the possible definitions based upon quantum mechanical considerations.

Recapitulating the foregoing discussion, it is clearly not our opinion that solution experiments appear to be inadequate for the purpose of comparison with molecular theories. However, we want to point out that due to the evident shortcomings of present theoretical and computational facilities a distinct scepticism is necessary in order to avoid the production of meaningless data. Of course, the solution experiments remain the main source of information, the data of which must be explained by theory. At the present stage of knowledge it is only possible to pick out selected properties of solutions which can be described satisfactorily on a molecular basis. For instance, referring to Frank and Wen's model of solvation shells [1], the structure of the inner shell should not be modified too much by interactions with the surrounding fluid, at least in the case of very strong solvation. Under such circumstances, the structure of a complex built up in the gas phase from the ion and the molecules yielding the inner solvation shell alone, would then represent an appropriate model of region a in Fig. 1, since interference from the bulk liquid should be damped down by the highly disordered outer solvent shell b. In turn such a gase-phase complex could be well mimicked by quantum mechanical model calculations.

Present quantum chemistry is mainly concerned with the calculation of bond energies. Accordingly, for the rest of this chapter we shall concentrate mainly on solvation enthalpies. The latter, being magnitudes composed of a lot of strongly interwoven effects, do not allow a simple separation of bulk and outer-shell influences.

Let us first recall that theoretical calculations on interactions between individual molecules yield energy changes for reactions of the types.

$$A^{\pm} + nS \longrightarrow AS_n^{\pm} \tag{38}$$

or

$$A^{\pm} + S_n \longrightarrow AS_n^{\pm} \tag{39}$$

where $A^{\pm}$ is the solute cation or anion, S is the solvent molecule, and n is a rather small number (frequently we have $n = 1$). Clearly processes (38) or (39) represent individual steps of the (imaginary) total single-ion solvation process [Eq. (39), $n \to \infty$], providing an appropriate thermodynamic cycle is built up. Eq. (38) describes an idealized solvation reaction in the vapor phase and provides a convenient basis for comparison with molecular calculation.

Owing to the advent of powerful techniques for the investigation on ion–molecule reactions [116,117] processes like (38) can in fact be directly measured in the gas phase. The direct observation of these reactions has, among others, the inherent advantage over solution experiments that no artificial separation into contributions of cations and anions needs to be performed, and that only well-defined supermolecular entities participate. Reaction (38) may be replaced by a series of consecutive steps to build up the cluster-ion $AS_n^{\pm}$:

$$A^{\pm} + S \longrightarrow AS^{\pm} \tag{40a}$$

$$AS^{\pm} + S \longrightarrow AS_2^{\pm}$$
$$\cdot \tag{40b}$$
$$\cdot$$
$$\cdot$$
$$AS_{n-1}^{\pm} + S \longrightarrow AS_n^{\pm} \tag{40c}$$

and accordingly $\Delta H_{n-1,n}^0$ would be the heat of reaction associated with step (40c). Similarly, the single solvent molecules required for the stepwise clustering can be obtained from the liquid by a vaporization process molecule by molecule.

$$S_{n+1} \to S_n + S \tag{41}$$

ΔH_n^V would be the heat of reaction of (41). Note that ΔH_n^V is dependent on n and when n is small it differs from the conventional heat of vaporization. Neglecting surface effects, the whole single-ion solvation processes can be constructed from reactions (40c) and (41)

$$
\begin{array}{lll}
S_{N-n+1} & \rightarrow S_{N-n}+S & \Delta H = -\Delta H_{N-n}^V \\
AS_{n-1}^{\pm}+S \rightarrow AS_n^{\pm} & \Delta H = \Delta H_{n-1,\,n}^0; \; n=1,\ldots,N \\
\hline
A^{\pm}+S_N \;\; \rightarrow A^{\pm}S_N = A_{Solv} & \Delta H = \Delta H_s^0 & (42)
\end{array}
$$

The total heat of solvation, ΔH_s^0, becomes

$$
\Delta H_s^0 = \sum_{n=1}^{N} (\Delta H_{n-1,\,n}^0 - \Delta H_n^V) \tag{43}
$$

Strictly, we require $N = \infty$, but the individual terms on the right-hand side of Eq. (43) will converge towards 0 as both $\Delta H_{n-1,\,n}^0$ and ΔH_n^V approach ΔH^V, the heat of vaporization of the solvent. Whereas (43) provides a "thermodynamic equation" for a molecular description of the solvation enthalpy, it is obvious that limits of both experimental and theoretical techniques will prevent its full use. Nevertheless, one would expect that the knowledge of the series of $\Delta H_{n-1,\,n}^0$ up to, say, $n = 9$ (at present the highest value observed in experiments) would allow for a crude extrapolation to large values of n and consequently for some qualitative or even semiquantitative predictions of relative values of ΔH_s^0 values for different ions. Today the challenge for a theoretical discussion of solvation energies is provided by these series of data (or, in connection with phenomenological descriptions, by the use of clusters rather than naked ions in proper Born-cycles [7,118]). Full agreement between estimated theoretical values and experimental single-ion solvation enthalpies should not be expected. However, data for gas-phase solvation, characterized by small numbers of solvent molecules must be reproduced by a serious theoretical calculation.

Clustering enthalpies are now available for a number of systems. A brief account of published data is given in the following section.

A. Gas-Phase Solvation

The important and stimulating contributions of Kebarle and co-workers [119–143] provide most of the data on gas-phase solvation. Several kinds of high pressure mass spectrometers have been constructed, using α-particles [121], proton- [123], and electron beams [144] or thermionic sources [128] as primary high-pressure ion sources. Once the solute $A^{\pm}$ has been produced in the reaction chamber in the presence of solvent vapor (in the torr region), it starts to react with the solvent molecules to yield clusters of different sizes. The equilibrium concentrations of the clusters are reached within a short time, depending on the kinetic data for the system. The cluster ions are then recorded in the mass spectrometer section of the apparatus. A detailed description of the experimental arrangement is given by Kebarle [145].

If equilibrium assumptions hold, the ratio of recorded ion intensities is equal to the ratio of the equilibrium concentrations of the cluster ions,

$$I_n/I_{n-1} = AS_n^\pm/AS_{n-1}^\pm \tag{44}$$

Thermodynamic data are easily available from the simple relations (45) and (46),

$$K_{n-1,n} = AS_n^\pm/AS_{n-1}^\pm \cdot P_s = I_n/I_{n-1} \cdot P_s \tag{45}$$

$$\Delta G_{n-1,n}^0 = RT \ln K_{n-1,n} \tag{46}$$

P_s being the total pressure of the solvent in the reaction chamber. Values for $\Delta H_{n-1,n}^0$ and $\Delta S_{n-1,n}^0$ can eventually be obtained from the temperature dependence of $\Delta G_{n-1,n}^0$. Known values for the individual steps $(n-1, n)$ can be combined to give the thermodynamic functions for any reaction (m, n). The total build-up of a solvated complex, characterized by $m = 0$, is of particular interest and corresponds to Eq. (38).

Experiments of the type described above have been conducted for a number of systems including, e.g. the ions H^+ and OH^-, the alkali- and halide ions, and the solvent molecules H_2O, NH_3, CH_3CN, CH_3OH, CH_3OCH_3, and others.

Tandem mass spectrometry [116,146] both stationary [116] and flowing afterglow-methods [116,147] and drift tube techniques [116] have also been applied to some of the clustering reactions. Results for the gas-phase solvation of H^+ by H_2O and NH_3 generally agree well with the values obtained by high pressure mass spectrometric observations [148].

Systems which might be of interest for molecular solvation, and which have been investigated by various techniques are indicated in Table 7. Gas-phase hydration enthalpies for the ions Pb^+ [158] and Bi^+ [159] are also given in Table 7. In their studies, Tang and Castleman [158,159] used an apparatus similar to that used by Kebarle and co-workers. Tantalizing as the existence of data for these ions might be for quantum chemists, who would prefer to know more about small ions like Be^{++}, it allows nevertheless the optimistic conclusion that a remarkable increase of activities in the field of gas-phase solvation can be expected in the near future. One should bear in mind that the experimental techniques were introduced only a few years ago. Probably very soon theoreticians will have at their disposal experimental reference data for a lot of interesting systems.

We shall now turn our attention to the results obtained for the hydration of alkali- [128,130,131,158] and halide ions [135,131,133,139], which have been the focus of all kinds of theoretical calculations. Some features are clearly evident from the gas-phase solvation data (Table 8).

1. Similar to the total single hydration enthalpies, the individual magnitudes $\Delta H_{n-1,n}^0$ decrease with increasing ion size in a given group.

Table 7. Survey of experimental literature pertinent to thermodynamical data for gas-phase solvation

Ion	H_2O	CH_3OH	$(CH_3)_2O$	NH_3	RNH_2[1]	HX[2]	YH[3]	CH_3CN
H^+[4,5]	118–120,123,124, 129,133,137,141, 142,148–154)	123,141, 142)	141,142)	121,122, 127,140, 148,155)	136,138, 156)			
H_3O^+	See H+			Not stable 140)				
NH_4^+[5]	140,148)			See H+				
Li^+	130)							
Na^+	130,158)							
K^+	128,130)							157)
Rb^+	130)							
Cs^+	130)							
Pb^+	158)							
Bi^+	159)							
O_2^-	125,132,139)	139)				132)		139)
OH^-	132,160)					132)		
F^-	135,131)							135)
Cl^-	125,131,134,139)	134,139)				143)	134)	135,139)
Br^-	125,131)							135)
I^-	125,131)							135)
NO_2^-	125)							

[1] $R = CH_3$ [138,156], C_2H_5, $n-C_3H_7$, $i-C_3H_7$ [156], C_6H_5 [136]. Moreover, $(CH_3)_2$, $(CH_3)_3N$, and mixed systems are reported [138].
[2] $X = F,Cl,Br,I$.
[3] $Y = (CH_3)_3CO$, C_6H_5O, $HCOO$, CH_3COO, CCl_3.
[4] Gas-phase proton affinity measurements are not included unless further clustering steps are also reported.
[5] Data for mixed solvent systems NH_3/H_2O, CH_3OH/H_2O, $(CH_3)_2O/H_2O$ and $CH_3OH/(CH_3)_2O$ are included in both respective columns.

Table 8. Gas-phase hydration enthalpies for individual clustering steps, $\Delta H^o_{n-1,n}$ and total single-ion hydration enthalpies ΔH^o_s for H+, alkali, and halide ions

Ion	$-\Delta H^o_s$[1]	$-\Delta H^o_{n-1,n}$							Ref.
		0,1	1,2	2,3	3,4	4,5	5,6	6,7	
H^+	269.8	165.0	31.6	19.5	17.5	15.3	13.0	11.7	124,137)
Li^+	133.5	34.0	25.8	20.7	16.4	13.9	12.1	—	130)
Na^+	106.1	24.0	19.8	15.8	13.8	12.3	10.7	—	130)
					13.6	11.6			158)
K^+	86.1	17.9	16.1	13.2	11.8	10.7	10.0		130)
Rb^+	81.0	15.9	13.6	12.2	11.2	10.5	—	—	130)
Cs^+	75.2	13.7	12.5	11.2	10.6	—	—	—	130)
F^-	113.3	23.3	16.6	13.7	13.5	13.2	—	—	131)
Cl^-	81.3	13.1	12.7	11.7	11.1	—	—	—	131)
Br^-	77.9	12.6	12.3	11.5	10.9	—	—	—	131)
I^-	64.1	10.2	9.8	9.4	—	—	—	—	131)

[1] See e.g. Ref. [27].

2. For a given ion the $\Delta H^0_{n-1,n}$ decrease for consecutive steps. In most cases[o] this effect becomes significantly smaller for a greater n. For the largest ions, Cs^+ and I^-, the effect of levelling off appears at remarkably small numbers of n. As has been pointed out above, we should expect this phenomenon of convergence for large cluster sizes, when solvent molecules added to the periphery no longer feel the specifity of the central ion. The asymptotic value for $\Delta H^0_{n-1,n}$ should be close to the value of the heat of vaporization of the solvent.

3. For the smallest ions, Li^+, Na^+, and F^-, particularly large values for the addition of the first and second water molecules are found.

Contrary to hydration, where no specific solvation shell structure around the halide ions is indicated by the continuously and smoothly decreasing enthalpies for subsequent clustering steps, acetonitrile reveals a particular solvation shell structure [139]. For $n = 1$ to $n = 3$ or 4, the $\Delta H^0_{n-1,n}$ are similar for H_2O and CH_3CN. For greater values of n a drastic decrease is noticed for acetonitrile [139]. The small extra stabilization brought about by the addition of a fourth molecule of CH_3CN to $Cl(CH_3CN)_3^-$ reflects the fact that no extended H-bonded network of solvent molecules can be formed, and that only a few molecules of solvent are tightly attached to the ion. When the space around the ion is densely packed, obviously by 3 CH_3CN-molecules, further solvent molecules have to take a peripheral position far from the ion, resulting in a reduced coulomb interaction. Similar effects are also found in the respective solvation of H^+ by water, methanol, and dimethyl ether [141]. While the proton affinities decrease in the order $CH_3OCH_3 > CH_3OH > H_2O$ [141] due to the electron donating properties of the CH_3-group, a switch-over is observed in the $\Delta H^0_{n-1,n}$ values for small n (Table 9).

All the phenomena summarized above are, at least for the smaller ions, to some extent accessible to a rather accurate quantum mechanical treatment and have therefore been presented in some length here. Much more stimulating information may be extracted from the gas-phase solvation experiments concerning features which are at present unsuitable for accurate quantum mechanical treatment but could be the basis for more approximate calculations or semiquantitative theoretical discussions. Along this line, we would like to point out two interesting observations. For the solvation complexes of H^+ with NH_3 there is strong evidence for a particularly stable entity $NH_4(NH_3)_4^+$, indicated again by a sharp drop of $\Delta H^0_{n-1,n}$ [140] (Table 10). For $n > 4$ solvent molecules would thus go into an outer solvation sphere with distinctly weaker bonding than in the first shell (see also [163]). The difference of 5 kcal/mole between $\Delta H^0_{3,4}$ and $\Delta H^0_{4,5}$ is certainly outside the range of experimental error. It is, however, a matter of personal opinion whether

---

[o] For F^-, a very sharp drop of $\Delta H^0_{n-1,n}$ is reported for $n = 2$, together with nearly identical values for $n = 3$, 4, 5 [131,145]. Although a pronounced fall might be expected for this particular system due to nonadditivity effects for hydrogen bonds [89], we do not regard this to be a fully satisfactory explanation for all the observations. Very accurate ab $initio$ calculations, which in general agree quite well with experimental results (see Chapter V), do not reproduce as sharp a drop in $\Delta E_{n-1,n}$ values for the hydrates of F^- [161,162]. Additionally, one should not forget that experimental errors of probably 10%, as quoted (in a different connection) by Grimsrud and Kebarle [141] could well account for the irregularities.

Table 9. Comparison of experimental gas-phase solvation enthalpies for the ions H^+, F^-, and Cl^- by different solvents.

Ion	Solvent	$-\Delta H^0_{n-1,n}$ (kcal/mole)					Ref.
		0,1	1,2	2,3	3,4	4,5	
H^+	H_2O	165.0	31.6	19.5	17.5	15.3	124,137)
	CH_3OH	182.0	33.1	21.3	16.1	13.5	141)
	$(CH_3)_2O$	187.0	30.7	10.1	—	—	141)
F^-	H_2O	23.3	16.6	13.7	13.5	13.2	131)
	CH_3CN	16.0	12.9	11.7	10.4	5.3	135)
Cl^-	H_2O	13.1	12.7	11.7	11.1	—	131)
	CH_3OH	14.2	13.0	12.3	11.2	10.5	139)
	C_6H_5OH	19.4	18.5	—	—	—	134)
	CH_3CN	13.4	12.2	10.6	6.2	—	135)
	$CHCl_3$	15.2	—	—	—	—	134)
	HCl	23.7	15.2	11.7	10.3	—	143)

Table 10. Thermodynamic data for individual clustering steps in gas-phase solvation of H^+, derived by high pressure mass spectrometry

Solvent	Quantity X	$-\Delta X^0_{n-1,n}$ kcal/mole					Ref.
		1,2	2,3	3,4	4,5	5,6	
NH_3	H	24.8	17.5	13.8	12.5	7.5	140)
	G	17.1	8.9	6.1	3.7	0.2	
H_2O	H	31.6	19.5	17.5	15.3	13.0	137)
	G	24.3	13.0	9.3	5.5	3.9	
CH_3OH	H	33.1	21.3	16.1	13.5	12.5	141)
	G	24.0	12.9	7.5	4.9	3.2	

one would take the small effects shown in the $\Delta G^0_{n-1,n}$ plots for the solvation of H^+ by H_2O and CH_3OH [141] as evidence for preferred shell structures or not. Here the difference could also be accounted for by experimental error.

Another result of considerable interest is the fact that the total single ion hydration enthalpies ΔH^0_s are nicely paralleled by the values for individual clustering steps in the gas phase, and hence even more markedly by the $\Delta H^0_{0,n}$ values (Figs. 7 and 8). It is not at all trivial that properties brought about by an infinite number of solvent molecules should be displayed already by such small complexes. Nevertheless, the shapes of the plots for ΔH^0_s and $\Delta H^0_{0,n}$ are quite similar.

Finally we note that all the aspects of gas-phase clustering and the relations to solvation phenomena have been discussed in detail by Kebarle in a recent authoritative review article [145].

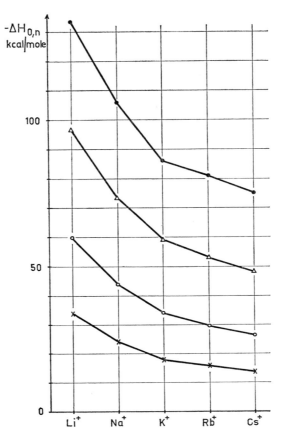

Fig. 7. Comparison of total single-ion solvation enthalpies ΔH_s^o ($\cdot$) and gas-phase solvation enthalpies $\Delta H_{o \cdot n}^o$, $n = 1$ ($\times$), $n = 2$(o), and $n = 4$($\triangle$) for the alkali ion series [130]

B. Ions in Solution

To round off the discussion of solvation enthalpies, reference is made here to the articles by Case [38] and by Friedman and Krishnan [27], who have reviewed the thermodynamic aspects of ion solvation extensively.

Gas-phase solvation has so far given only very indirect evidence concerning the structure and details of molecular interactions in solvation complexes. Complex geometries and force constants, which are frequently subjects of theoretical calculations, must therefore be compared with solution properties, however, the relevant results are obscured by influences arising from changes in the bulk liquid or by the dynamic nature of the solvation shells. With few exceptions, structural information from solutions cannot be adequately resolved to yield more than a semiquantitative picture of individual molecular interactions. The concepts used to convert the complex experimental results to information for structural models are often those of solvation numbers [33], and of structure-making or structure-

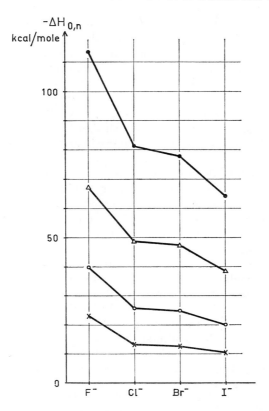

Fig. 8. Comparison of total single-ion solvation enthalpies ΔH_s^o ($\bullet$) and gas-phase solvation enthalpies $\Delta H_{0,n}^o$, $n = 1(\times)$, $n = 2(o)$, and $n = 4$ ($\triangle$), for the halogenid ion series [131]

breaking properties [1] indicating whether the degree of structural order of the liquid solvent is enhanced or decreased by dissolved ions.

Several methods are successfully applicable in this field, *e.g.* dielectric relaxation methods [164], IR investigations in the near, fundamental, and far IR regions [165], RAMAN spectroscopy [166], NMR spectroscopy [32,34,167], and ultrasonic absorption [168-170].

Specific data are obtainable in only a few cases. A number of cations, including Mg^{2+}, Al^{3+}, and others [171] are so strongly hydrated, that a rather rigid solvent cage is built up around the ion. At low temperatures the exchange rate of molecules in this tightly bound state in the bulk liquid is slow enough to give well separated signals in NMR measurements, allowing a direct determination of the number of water molecules in the primary hydration shell [171]. For Be^{2+}, the coordination number turns out to be 4, for the other cations concerned the solvation number of 6, which is in this case directly related to a geometrical arrangement, tells us that a more or less distorted octahedron of solvent molecules forms the inner solvation shell [171].

Some recent papers permit an exciting outlook on the degree of sophistication of experimental techniques and on the kind of data which may be available soon. In the field of NMR spectroscopy, a publication by Hertz and Raedle [172] deals with the hydration shell of the fluoride ion. From nuclear magnetic relaxation rates of ^{19}F in 1 M aqueous solutions of KF at room temperature, the authors were able to show that the orientation of the water molecules in the vicinity of fluoride ions is such that the two protons are non-equivalent. A geometry is proposed for the water coordination in the inner solvent shell of F^- corresponding to an almost linear H-bond and to an OF distance of approximately 2.76 Å, at least under the conditions chosen.

Another note shall be dedicated to X-ray diffraction investigations of electrolyte solutions, which have now already 45 years of history. In principle, diffraction patterns contain direct information of average arrangements and geometrical parameters of the liquid solutions examined, but these patterns result from a superposition of all kinds of neighbor-relations existing in solutions. The conversion of the one-dimensional data obtained in the experiments to a three-dimensional geometrical model of the solution is neither straightforward nor unique, being a procedure of trial and error. In particular, water molecules are seen as nearly unstructured spheres by X-rays. Consequently, information concerning orientational correlations are completely lacking. By also considering concentration dependence and by starting from plausible models — supported by different techniques — the non-uniqueness may be reduced to some extent. Bertagnolli et al. [173] report investigations on CsF solutions following this philosophy. They performed X-ray diffraction studies of concentrated aqueous solutions of CsF at room temperature corresponding to compositions from saturation ($\sim CsF.2.3H_2O$) down to $CsF.8H_2O$, and find their data compatible only with a model of correlation clusters, inside which the structure is determined by a β-tridymite-like correlation lattice. The Cs^+ ions are placed in the cavities of the lattice. Clearly, no single ion properties can be derived from these results due to the occurrence of strong ionic interactions and the absence of the bulk liquid in concentrated solutions.

A most appealing strategy for structural investigations is a combination of X-ray and neutron diffraction methods, which are somehow complementary to each other. Neutrons, being scattered distinctly by hydrogen nuclei, can be used to resolve both orientational correlations in the liquid phase and internal geometries of the water molecules. Since the first combined application of both techniques by Narten [174], who investigated liquid water, aqueous solutions of LiCl [175] have been examined at room temperature over a wide range of concentrations up to saturation ($\sim LiCl.3H_2O$). Even rather dilute solutions corresponding to a molar ratio $H_2O:LiCl$ of 166 have been considered. The results allow for a rather detailed analysis of the structure of aqueous LiCl solutions, starting from an assumed model for water as derived by Narten's measurements [174]. They can be regarded to comprise the most complete structural description of an electrolyte solution known at present, and are as follows [175]: In sufficiently dilute solutions, there are three main regions, one of them is characteristic of pure water in which tetrahedrally coordinated water clusters are predominant, and in which the geometry of the water molecules themselves is, on the average, close to vapor phase values. This basic water structure disappears when the molar ratio is 10 or less. The other two regions may be ascribed to the hydrated ions. Li^+ appears to be tetrahedrally coordinated by four water molecules, a lone pair of oxygen pointing toward the cation. Cl^- has an average coordination number of 6, the water molecules are linearly hydrogen bonded to the anion and prefer an octahedral shell. These structures also prevail in highly concentrated solutions. For all but very concentrated LiCl solutions, dynamic effects cause the coordination numbers to fluctuate sufficiently to be given as 4 ± 1 and 6 ± 1 respectively. The coordination distances between ions and water molecules vary with concentration. Obtained RMS variations for all distances are large and indicate, that at any instant a considerable fraction of the molecules is present in largely distorted internal and intermolecular configurations.

These results of Narten et al. [175] are an excellent basis for a recapitulation of the preceding short account on the status of experiments. We shall first derive a picture of ionic solutions and the connection lines to gas-phase solvation and to molecular theories in a speculative manner, which must not be taken really literally, but should be considered to be an artificially optimistic stimulus for a concerted effort of theory and experiment.

First of all, diffraction data are interpretable without discrepancies using a solvation model like the one of Frank and Wen [1]. (However, this statement should not be understood as a formal proof for the indicated model, since at low concentrations, where it should work best, an interpretation of the small effects observed in the diffraction patterns might be quite insensitive to the model applied.)

Let us now start from a very concentrated solution according to LiCl. $3H_2O$, which is similar in composition to the crystalline dihydrate precipating from a saturated LiCl solution. We might therefore also expect to find some structural similarities between the crystalline hydrate and the solution. In the latter, well-oriented layers of water molecules are observed around both Li^+ and Cl^-, tetrahedrally coordinated ones at the cation, and an octahedral one around the anion. The average orientations of the molecules are already similar to the most stable gas-phase structures displaying a lone-pair interaction of the water oxygen with Li^+ and a hydrogen bond towards Cl^-. However, there is only a very small total number of water molecules, and many of them have to be shared between the hydration layers of two ions. This places a serious geometrical constraint on the structure as a whole and a compromise between the competing interactions is necessary. The situation does not remind so much one of solvated ions but is reminiscent of a crystal. Indeed the average coordination distances are characteristic for crystalline hydrates [176]. The OCl distance is just the average value observed in crystals ($R_{OCl} = 3.19$ Å), and the OLi value is particularly high here ($R_{OLi} = 2.25$ Å). The liquid character of the solution is displayed by the high fluctuation of these values (RMS ~ 0.25 Å).

In more and more dilute solutions, added water molecules would lower the strain as these molecules would be taken up into the hydration shells therefore allowing a progressive separation of the hydrated ions. When the molar ratio falls to 8, the average distances decrease ($R_{OCl} = 3.10$ Å, $R_{OLi} = 1.94$ Å). Still, the instantaneous deviations from these values are high. As the molar composition of the solution becomes about 10, enough water molecules are present to guarantee each structural unit, $Cl^-(H_2O)_6$ or $Li^+(OH_2)_4$, its individual existence. The diffraction pattern commences to be shifted towards the one of pure water, indicating that the excess water molecules are present in a state comparable to the one of pure water, characterized by clusters of tetrahedral coordination [174]. In addition to the three different types of clusters, *i.e.* pure water, hexahydrated chloride, and tetrahydrated lithium cation, a small amount of water is visualized in an approximatively random configuration and is highly mobile (the region *b* in the Frank and Wen-model [1]).

At the same time, as the concentration decreases the exchange of water molecules by cooperative processes becomes easier and so significant fluctuations in the coordination numbers are observed, which are estimated to be about ± 1. At even lower concentrations, ion–water correlation patterns will become obscured and then undetectable. As an extrapolation, dynamic processes might contribute more and more to the description of the solution. The strong interaction between Li^+ and OH_2 ($\Delta H_{0,1}^0 = 34$ kcal/mole in the vapor phase [130]) may cause the cation–water complexes to remain quite well-defined tetrahydrates, on the average, despite all dynamic effects.

Obviously this picture might be supported and supplemented by according data from different experimental investigations, or it might be modified to fit these data. Interactions within the basic hydrated structures, as well as their energetics, are obtainable from gas-phase solvation experiments or from accurate MO calculations. For the simulation of real solutions, dynamic calculations will be inevitable. There is, however, a demand for acceptable effective potentials to be used in molecular dynamics, or in Monte Carlo calculations.

In conclusion of this short account on experiments, which is clearly far from complete, detailed structural data for solutions will be available in the near future. They may serve well to support theoretical calculations of solvation processes and to present challenges for theoretical considerations, which will in any case have to be dynamic ones. Data which may be compared quantitatively with molecular calculations will, however, have to come from gas-phase solvation experiments. There already exists a great variety of according data and their number will certainly increase further.

IV. Electrostatic Models and Further Developments

A. Electrostatic Models

One of the first successful approaches to a structural theory of ionic solvation was the use of classical electrostatic models [4]. The concept is to choose a structural model thought to be representative for the solvated species, to split the energy of interaction into several attractive and repulsive parts, each of which should have an easily interpretable physical meaning, and to minimize the energy with respect to the complex geometry. Another energy is obtained when a solvent molecule is placed into the cage instead of the ion. The difference between these two numbers plus the energy of condensing a solvent molecule to the bulk is then regarded as an approximation of the solvation energy. Further refinements are frequently made by incorporating the interaction of the complex as a whole with the bulk solvent in the framework of a conventional continuum theory [4,6]; effects due to volume relaxation are only rarely taken into account [6].

Evidently the freedom of selecting a "representative" model and the arbitrariness of choosing analytical functions for the partitioned potential led to the invention of a large variety of closely related models, Bernal and Fowler's work [4] is one of the earliest attempts in this direction. A large amount of similar work has been done in this field: Eley and Evans [177] discussed heat and entropy changes during the solvation process for tetrahedrally coordinated ions using essentially the same ideas as Bernal and Fowler. Muirhead-Gould and Laidler [178] generalized the model for the inclusion of higher multipole interaction and arbitrary coordination numbers. Verwey [179] arrived at a model where the ion is relaxed from lying in a common plane with the ligand, thus achieving further stabilization by interaction of ligands with molecules of the subsequent layers.

Further details concerning the older theories can be found in summaries [26,38]. More recent work of Spears [105] deals with the tricky problem of choosing a suitable set of repulsion parameters for the atoms involved. In their recent paper, Eliezer and Krindel [180] account more carefully for polarization effects than is usually done. The most elaborate approach up until now has been undertaken by Morf and Simon [6,181]. They split the free energy of solvation into the following terms:

1. The generation of a suitable cavity in the solvent for the inclusion of the ion.

2. The energy of interaction of the ion with water molecules of the first hydration layer as well as the interaction of these molecules among themselves.

3. The influence the ion exerts on the bulk of the solvent, including volume dilation.

4. The compilation of the free energy values corresponding to the tabulated standard concentrations[178].

The interaction in the coordination sphere is described by a 13-term function considering multipole-, polarization-, dispersion-, and repulsion interactions. They succeeded in reproducing free energies of solvation of a large variety of cations to within standard deviations of a few percent. This formalism has been applied by the authors to a discussion of the influence of the several parameters of the model on ion-selectivities in certain types of ionophores [181].

The use of molecular electrostatic potentials calculated from the wave function of ligands, as has been recently suggested [50-54] seems to present a less ambiguous alternative to the more empirical approach described above. This potential can either be calculated exactly for each point of interest according to Eq. (3), or it is approximated by a suitable distribution of point charges choosen either by intuitive guess or by a less arbitrary method like the one of Hall [182].

At the present stage of development this method cannot account for terms like polarization, dispersion, charge transfer, or repulsion energy. Nevertheless it was found to be appropriate for the prediction of preferred solvation sites of polyatomic molecules [50-54]. Figure 9 illustrates the use of the method in ion–water interactions. The plot shows the variation of potential energy of a negative unit charge rotating around a water molecule at different distances. The corresponding plots for a positive charge is obtained simply by changing the sign on the energy axis. Since only the charge of the ion enters the expression for the electrostatic energy, no distinction is made between different ions of the same charge. Nevertheless, the topology of the complexes is predicted correctly for spherically symmetric ions. Only one energy minimum is found for cation–molecule complexes, representing the structure with C_{2v}-symmetry. Two equivalent hydrogen-bonded positions are favored for anions at small distances which gradually coincide as the distance increases to give again the C_{2v} structure determined by ion-dipole interaction. As the use and the general features of the molecular electrostatic potential have been described extensively in a review paper in this series [19] we will not go into further detail here but only mention more recent attempts to compare directly the electrostatic and the full H. F. formalism in several examples like the hydration of formamide [50], N-methylacetamide [54], purines, and pyrimidines [52]. The smaller the solute–solvent distance is, of course, the greater are the deviations from the Hartree-Fock results. In case highly polarizable regions are present in one of the constituents of the complex, the model assumptions are violated and deviations become noticable. To give an example, hydration around the carbonyl group of formamide with highly polarizable π-electrons is reproduced rather poorly [50].

One of the greatest advantages of the partitioned potential methods is their conceptual simplicity, which permits rapid evaluation of the energy of a given configuration. This feature is an essential prerequisite for more elaborate dynamic treatments of solvation [11-13,183]. Despite their simplicity these theories are unsatisfactory in many ways. Their applicability is limited to structures in which the changes in the individual components on complex formation are not too drastic. Wrong geometries are sometimes predicted for ions capable of forming hydrogen bonds [95,130,180]. Apart from the mostly ambiguous choice of how to split the energy and what special kind of potential function to employ, some of the energy contributions cannot be accounted for in a transparent way, charge-transfer energy serves as an example. Furthermore, the single contributions are intercorrelated rather than separable as the partitioned potential method implies. If for example a multipole moment is induced in one of the components of the complex, it reacts back not only to influence the charge distribution that generated it,

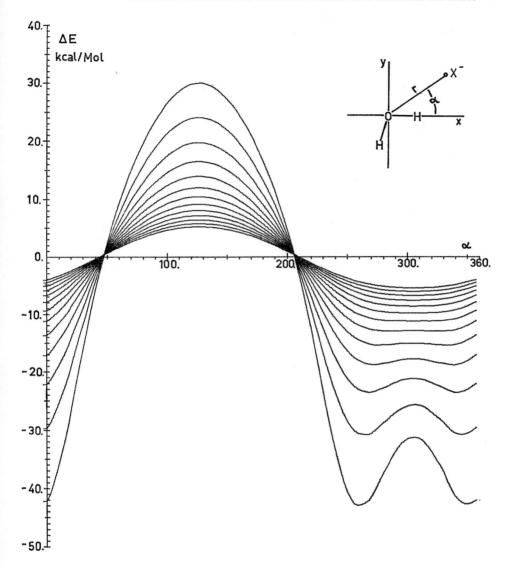

Fig. 9. Potential of a negative unit charge rotating around a water molecule at different distances r [$r = 4.0$–10.0 (a. u.) in steps of 0.5 (a. u.)]

but also on all other contributions like dispersion and repulsion energy in a very complex way.

Extension of the theory to some sort of self-consistency would on the other hand certainly lead to a formalism difficult to survey. Additionally more and more empirical parameters would enter the theory, thus destroying its formerly appealing simplicity.

B. Further Developments for the Treatment of Ion-Solvent Interactions

Finally some other developments are briefly mentioned for the sake of completeness. These include the work of Sposito and Babcock [184] which bears close resemblance to the partitioned potential methods. Their idea is to solve the quantum mechanical energy spectrum of the complex using only the empirical potential function as the potential operator in the Hamiltonian. This spectrum then leads to calculations of temperature-dependent energies of formation.

If one is interested in changes of the solute molecule, or if the structure of the surrounding solvent can be neglected, it may be sufficient to regard the solvent as a homogeneous dielectric medium, as was done in the older continuum theories, and to perform a quantum mechanical calculation on the molecule with a modified Hamiltonian which accounts for the influence of the solvent as has been done by Hylton et al. [185]. Similarly Yamabe et al. [186] substituted dipole-moment operators for the solvent in their perturbational treatment of solvent effects on the activation energy in the $NH_3 + HF$ reaction.

Another attempt was made by Klopman [187] who proposed to simulate the effect of the solvent by a set of auxillary particles he called "solvatons"; his concept was extended later by Germer [188]. Methods for the discussion of changes in spectral properties have been summarized by Basu [189] and Liptay [190]. These methods could perhaps lead to discussion of some solvent effect on larger systems but regrettably no definite conclusions on solvent structure are possible. Starting from electrostatic pair potentials Heinzinger and Vogel [183] performed molecular dynamic studies on aqueous LiCl solutions. For the purpose of comparison with similar calculations using *ab initio* potentials, this paper is discussed in Chapter V.

V. Results from Quantum Chemical Calculations

In order to gain deeper insight into the nature of ion–molecule interactions it is necessary to rely upon quantum mechanics. Thereby we can obtain, at least in principle, a description of the observable phenomena (including those nonaccessible by classical theories) which is precise and free of parameters. In recent years considerable interest has evolved in calculations of this type.

For obvious reasons water is the most thoroughly treated solvent molecule. It is small enough to be calculated with high accuracy, it is capable of forming hydrogen bonds, and its significance in nearly all branches of chemistry and biology need not be mentioned here in detail. In the second place the components of the ester and amide groups have received special attention, which is mainly due to their occurrence in biologically important molecules like ionophores, peptides, proteins, and nucleic acids, as well as in pharmacologically active compounds. Ammonia as a solvent molecule has also been considered in a few examples. The cations of the alkaline and earth-alkaline series and the halogenid- ions up to the second or third row represent the best-suited examples for solvation studies both from the experimental and the computational point of view. Accurate investigations of larger molecular systems and heavy ions are prohibited for computational reasons until now. An approximate treatment of heavy ions was suggested recently [191–194].

A. Cation-Solvent Interactions

1. One to One Complexes

Although simple, a model system containing one solvent molecule together with one ion already provides valuable insight into the nature of the ion–solvent interaction. There is also convincing evidence that this two body potential dominates in much more complicated situations like in the liquid state [88,89,162]. Molecular data for one to one complexes can be calculated with sufficient accuracy within reasonable time limits. Gas-phase data reported in Chapter III provide a direct basis for comparison of the calculated results.

Some of the most important properties of ion–solvent interaction have been studied where water is used as the solvent molecule. Furthermore, these calculations have frequently been made with fairly extended basis sets. Therefore we draw our attention first to hydration studies.

a) Cation-Hydration Studies. Equilibrium geometries and energy minima for some typical monohydrated cations calculated in the last five years, together with some information concerning the basis sets are summarized in Table 11. All the calculations involving monatomic ions indicate the complexes to be planar and to show C_{2v} symmetry. In accordance with classical electrostatics, the cations face the negative end of the water dipole. The concept of two opposite effects dominating interaction energies at the equilibrium geometries, ion–dipole interaction and closed shell repulsion, is illustrated nicely by the two series of central ions with increasing charge and size, $Na^+ \cdot OH_2$, $Mg^{2+} \cdot OH_2$, $Al^{3+} \cdot OH_2$ and $Li^+ \cdot OH_2$, $Na^+ \cdot OH_2$, $K^+ \cdot OH_2$, respectively. The equilibrium geometries

Tabelle 11. *Ab initio* calculations on cation–water complexes

Cation	Basis set[1])			Ref.	$-\Delta E$ (kcal/mole)	R_{OX+} (Å)	Remarks
	Orbitals	uncontracted	contracted				
Li+	STO4G	—	—	195)	31.0	2.30	R_{OLi} not varied
Li+	GTO	O(9/5) H(3) Li(9/1)	[2/1] [1] [2/1]	111)	47.4	1.81	
Li+	GTO	O(9/5) H(3) Li(9/1)	[4/2] [3] [4/1]	111)	46.6	1.82	
Li+	GTO	O(9/5) H(4) Li(9/1)	[6/4] [4] [7/1]	111)	43.6	1.83	
Li+	GTO	O(9/5/1) H(4/1) Li(9/3)	[4/2/1] [2/1] [4/2]	90)	37.0	1.85	
Li+	GTO	O(11/7/1) H(6/1) Li(11/2)	[5/4/1] [3/1] [5/2]	196)	36.0	1.89	
Li+	GTO	O(11/7/1) H(6/1) Li(7/1)	[4/3/1] [2/1] [3/1]	87)	34.4	1.89	
Li+	GTO	O(13/8/2/1) H(6/2) Li(8/2/1)	[8/5/2/1] [4/2] [5/2/1]	87)	35.2	1.89	
Li+	GTO	O(9/5) H(4) Li(4)	[2/2] [2] [3]	194)	39.7	1.88	Pseudo-potential calculation
Li+	—	—	—	124,137)	34.0	—	Exp. vmlue
Be2+	GTO	O(9/5) H(4) Be(9/1)	[6/4] [4] [7/1]	111)	141.4	1.56	
Be2+	GTO	O(9/5/1) H(4/1) Be(9/3)	[4/2/1] [2/1] [4/2]	90)	140.0	1.50	
Na+	STO3G	—	—	197)	40.7	1.99	
Na+	GTO	O(9/5/1) H(4/1) Na(11/7)	[4/2/1] [2/1] [6/4]	90)	27.0	2.20	
Na+	GTO	O(11/7/1) H(6/1) Na(14/8/1)	[5/4/1] [3/1] [8/6/1]	196)	25.2	2.24	
Na+	GTO	O(11/7/1) H(6/1) Na(13/8/1)	[4/3/1] [2/1] [7/4/1]	198)	23.9	2.25	

48

Table 11 (continued)

Cation	Basis set[1)			Ref.	$-\Delta E$ (kcal/mole)	R_{OX}^+ (Å)	Remarks
	Orbitals	uncontracted	contracted				
Na+	GTO	O(9/5) H(4) Na(6)	[2/2] [2] [3]	194)	27.1	2.36	Pseudo-potential calculation
Na+	—	—	—	130,158)	24.0		Exp. value
Mg2+	GTO	O(9/5/1) H(4/1) Mg(11/7)	[4/2/1] [2/1] [6/4]	90)	80.0	1.95	
Ca2+	GTO	O(9/5/1) H(4/1) Ca(14/9)	[4/2/1] [2/1] [8/6]	90)	53.0	2.40	
Al3+	GTO	O(9/5/1) H(4/1) Al(11/7)	[4/2/1] [2/1] [6/4]	90)	180.0	1.75	
K+	STO3G	—	—	197)	27.9	2.40	
K+	GTO	O(9/5/1) H(4/1) K(14/9)	[4/2/1] [2/1] [8/6]	90)	18.0	2.65	
K+	GTO	O(11/7/1) H(6/1) K(17/11/1)	[4/3/1] [2/1] [11/7/1]	198)	16.64	2.69	
K+	GTO	O(9/5) H(4) K(7)	[2/2] [2] [4]	194)	18.3	2.90	Pseudo-potential calc.
K+	—	—	—	130)	17.9	—	Exp. value
NH$_4^+$	STO3G	—	—	199)	37.3	2.40	Linear H-bridge
					16.0	2.60	Along a C_{3v} axis
NH$_4^+$	STO4-3/1	O/N(8/4) H(4)	[4/2] [2]	199)	28.1	2.65	Linear H-bridge
NH$_4^+$	—	—	—	140)	17.3	—	Exp. value
CH$_3$N+H$_3$	STO3G	—	—	199,200)	34.2	2.45	Linear N–H...O H-bond
(CH$_3$)$_4$N+	STO3G	—	—	200)	5.4	CO 3.1	Along N-C axis
					6.4	NO 3.7	Bisecting two N–C bonds
					10.3	NO 3.4	Axial along C_{3v} axis
					10.0	—	Linear C–H..O H-bridge

Table 11 (continued)

Cation	Basis set [1]			Ref.	$-\Delta E$ (kcal/ mole)	R_{OX^+} (Å)	Remarks
	Orbitals	uncontracted	contracted				
EtN$^+$ (Met)$_3$	STO3G	—	—	200)	5.6	CO 3.1	Bisecting α-CH$_2$ angle
					3.4	CO 3.1	Coaxial with H$_3$C–CH$_2$ axis
					9.0	—	α C–H H-bonded
					6.0	—	β C–H H-bonded
(Et)$_2$N$^+$ (Met)	STO3G	—	—	200)	8.7	NO 3.6	30° deviation from N(Met)$_2$ angle bisectrix
(Et)$_3$N$^+$ Met	STO3G	—	—	200)	8.2	—	

[1] The notation GTO A($\alpha/\beta/\gamma/\delta$) [$a/b/c/d$] means: α s-type, β p-type, γ d-type, and δ f-type Gaussian Type Orbitals are centered on atom A which are contracted to a s-type, b p-type, c d-type, and d f-type orbitals respectively.
The notation STOnG means: a minimal atomic basis of Slater Type Orbitals is used each of which is expanded into n-simple Gaussian Type Orbitals.

of complexes formed by polyatomic ions, however, cannot be predicted in such a simple way. In the ammonium ion and related species [199,200] linear hydrogen bonds are formed and determine the structure of the complexes.

If only monovalent cations are considered, the results compare well with classical predictions, but a detailed analysis of the different contributions to the energies of interaction — discussed later in this chapter — shows that the agreement is rather fortuitous, because two contributions, the nonclassical exchange energy and the delocalization term including mainly polarization and charge transfer, are almost cancelled out in the vicinity of the energy minimum. If one of the two contributions happens to be anomalously high, as in Be$^{2+}\cdot$OH$_2$ [111], the classical scheme fails to reproduce the correct shape of the energy surface.

Considering the different calculated values for an individual complex in Table 11, it seems appropriate to comment on the accuracy achievable within the Hartree-Fock approximation, with respect to both the limitations inherent in the theory itself and also to the expense one is willing to invest into basis sets. Clearly the Hartree-Fock-Roothaan expectation values have a uniquely defined meaning only as long as a complete set of basis functions is used. In practice, however, one is forced to truncate the expansion of the wave function at a point demanded by the computing facilities available. Some sources of error introduced thereby, namely "ghost" effects and the inaccurate description of ligand properties, have already been discussed in Chapter II. Here we concentrate on the central ion: d-type polarization functions are indispensable for ions containing occupied p-orbitals. Otherwise the two orbitals oriented perpendicular to the axis joining ion and ligand cannot contribute to the polarization of the metal cation. As an example the SCF-polarizability calculated for a sodium cation without d-type functions is approximately the same as that for a lithium cation with p-type polarization functions, whereas the experimental value is about

Table 12. Calculated and experimental polarizabilities of some alkaline cations (values in Å^3)

Ion	H. F. calculation		Experimental value
Li+	0.030[1]	0.026[2]	0.027[3]
Na+	0.155[1]	0.026[2]	0.210[3]
K+	0.944[1]	—	0.870[3]

[1]) See Ref. [201].
[2]) Calculated without d-type functions.
[3]) See Ref. [202].

eight times as large (see Table 12). Full basis sets, on the other hand, guarantee a fairly correct description of ion polarizabilities. Two other simplifications frequently made are more likely to be justified: the first of them is the assumption that the solvent molecule does not appreciably change geometry during complex formation, an approximation which seems reasonable from chemical experience, the second one is the neglect of correlation energy.

Several authors have relaxed the rigidity constraint in order to test the first approximation. The results of full optimization are summarized in Table 13. In all cases investigated the corrections are within the error limits of H. F. calculations and justify the assumption of frozen ligand geometry. The second point, the neglect of correlation energy, is inherent in the H.F.-method (cf. Chapter II), and can only be removed if the single determinant used to approximate the wave function is replaced by a linear combination of different configurations. Since calculations including configuration interaction are both extremly time consuming and slowly convergent, this procedure has been used in only very few cases of ion–molecule complexes until now [108]. Moreover, in principle the process of complex formation can be described correctly within the H.F. approximation, because only closed shell entities are involved [58]. Therefore one is inclined to assume that the contribution of electron correlation to the energy of complex formation is rather small. An explicit calculation performed on the Li+–H_2O complex using a 5613 configuration wave function [108] destabilizes the complex only slightly ($\Delta E_{COR} = 1$ kcal/mole, see Chapter II). Other estimates of the correlation binding energy have been undertaken by Kistenmacher et al. [206] who used a statistical method of Wigner [207] to calculate the desired energy directly, rather than as a difference between two large numbers. The assumptions

Table 13. Effect of geometry optimization

Ion	Ref.	Basis set[1] Uncontracted	Contracted	H_2O geometry[2] unperturbed	Interaction energy[2] with rigid H_2O $-\Delta E$	Complex geometry[2] with H_2O geometry relaxed	Optimized[2] $-\Delta E$
Li^+	[87]	O(11/7/1) H(6/1) Li(7/1)	[4/3/1] [2/1] [3/1]	$R_{OH} = 0.950$ $\alpha_{HOH} = 106.6°$ (H.F.-minimum)	34.2 $R_{OLi^+} = 1.89$	$R_{HO} = 0.955$ $\alpha_{HOH} = 106.1°$ $R_{OLi^+} = 1.89$	34.20
Li^+	[196]	O(11/7/1) H(6/1) Li(11/2)	[5/4/1] [3/1] [5/2]	$R_{OH} = 0.958$ $\alpha_{HOH} = 104.52°$ (exp. geometry)	36.00 $R_{OLi^+} = 1.89$	$R_{OH} = 0.958$ $\alpha_{HOH} = 106.1°$ $R_{OLi^+} = 1.86$	36.00
F^-	[203]	F/O(11/7/4) H(6/1)	[5/4/1] [3/1]	$R_{OH} = 0.958$ $\alpha_{HOH} = 104.52°$ (exp. geometry)	24.10 $R_{OF^-} = 2.52$	$R_{OH_1} = 1.01$ $\alpha_{HOH} = 104.52°$ $R_{OF^-} = 2.25$	24.90
F^-	[204]	O(11/7/1) F(13/8/1) H(6/1)	[4/3/1] [7/3/1] [2/1]	$R_{OH} = 0.950$ $\alpha_{HOH} = 106.6°$ (H.F. minimum)	22.85 $R_{OF^-} = 2.56$ $\delta = 5.2°$	$R_{OH_1} = 0.995$ $R_{OH_2} = 0.946$ $\alpha_{HOH} = 102.7°$ $R_{OF^-} = 2.51$ $\delta = 4.5°$	24.42
Cl^-	[204]	O(11/7/1) Cl(17/10/1) H(6/1)	[4/3/1] [9/6/1] [2/1]	$R_{OH} = 0.950$ $\alpha_{HOH} = 106.6°$ (H.F. minimum)	11.93 $R_{OCl^-} = 3.31$ $\delta = 14.6°$	$R_{OH_1} = 0.961$ $R_{OH_2} = 0.946$ $\alpha_{HOH} = 102.43$ $R_{OCl^-} = 3.30$ $\delta = 14.6°$	12.10
Cl^-	[205]	double ξ STO	—	$R_{OH} = 0.975$ $\alpha_{HOH} = 108.6°$ (H.F. minimum)	18.40 $R_{OCl^-} = 3.04$	$R_{OH_1} = 1.02$ $R_{OCl^-} = 3.04$	19.00

1) GTO basis sets are used unless stated otherwise, for the notation of the basis see Table 11.
2) All lengths in (Å), all energies in (kcal/mole)
   H_1 denotes the bridge proton, for a definition of δ see Fig. 19c.

underlying this procedure, namely negligible differences between uncorrelated and correlated total electron densities, seem to be justified. They also conclude that the correlation correction is always sufficiently small to be neglected in this special type of ionic complexes (Table 14). This statement is not even affected by the fact that their estimate shows opposite sign to Roos' and Diercksen's result [108]. Much greater weight is to be attributed to effects of nuclear motion. These include vibrational and rotational contribution to the heats of formation. Usually these terms are neglected not only because of their assumed smallness, but also for the difficulties that arise in their calculation. Nevertheless estimates have been made to account at least for the most important vibrational corrections. Kistenmacher et al. [206] have treated this correction in an approximate way. They obtained contributions of zero point vibration to complex stability in the order of some kcal/mol (see Table 14).

Table 14. Smaller contributions to the calculated heats of formation for some ion–H_2O complexes[1])

	$Li^+...OH_2$	$Na^+..OH_2$	$K^+..OH_2$	$F^-..HOH$	$Cl^-..HOH$
H. F. binding energy	−35.20	−23.95	−16.64	−23.70	−11.85
Correlation correction	− 0.26	− 0.26	− 0.33	− 0.75	− 0.57
Zero point energy correction	2.24	1.74	1.51	3.40	1.80
Computed $-\Delta H_{298°K}$	34.13	23.32	16.20	22.18	11.37
Estimated uncertainty	± 0.94	± 1.00	± 0.80	± 0.91	± 1.70
Experimental $-\Delta H$	34.0	24.0	17.90	23.3	13.10

[1]) All values taken from Ref. [206]. Energies in (kcal/mole).

The corrections brought about by molecular vibrations suggest more detailed investigations of nuclear dynamics. It is perhaps interesting to mention here that the binding-energy of $Li^+\cdot OH_2$ is about 1.2 kcal/mole higher than that of $Li^+\cdot OD_2$ [105] an effect primarily due to the anharmonicity of the normal modes involving hydrogen motion.

The most vivid picture of the appearance of the energy surfaces of the complex is obtained by means of contour diagrams that show iso-energy curves in a given plane of the solvent molecule at a fixed position. Kistenmacher et al. [206] presented analytical fits to approximate the energy hypersurfaces. They intuitively identified the resulting plots as shapes of the water molecule as it is felt by the different ions. A study of these surfaces is very instructive in discussing mobilities of the ions in the field of the solvent molecule as well as steric effects. In addition to their instructive value they provide the starting information for more advanced techniques like molecular dynamics. Figs. 10a–10c show the

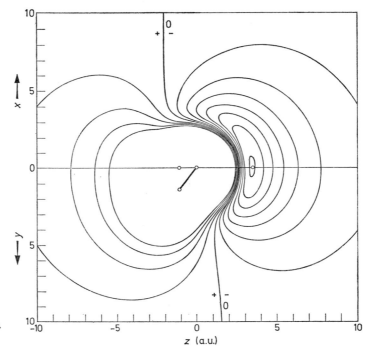

Fig. 10a

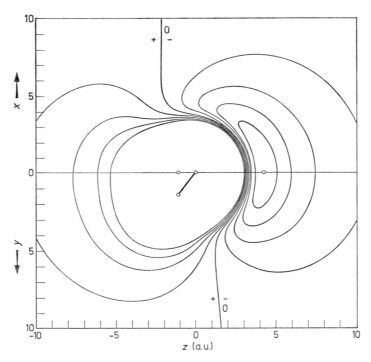

Fig. 10b

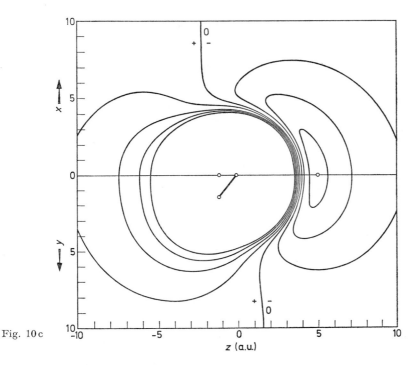

Fig. 10 c

Fig. 10. a–c. Sections through the energy hypersurfaces of the systems Li^+–OH_2, Na^+–OH_2 and K^+–OH_2. Lower part: section in the plane of the water molecule Upper part: section perpendicular to the plane of the water molecule containing the C_{2v}-axis. The spacing of the lines is 5.0 kcal/mole (analytical fit from Ref. [206])

resulting plots for planes normal to or coinciding with the plane of the water molecule. Interestingly, the surfaces for the three ions Li^+, Na^+, and K^+ look very much alike in the *repulsive* region. In the attractive part cations show less mobility in the plane of the molecule than perpendicular to it, a result which can be easily explained by the repulsive effect of the hydrogens. These details become less important as the size of the cations increases.

Additional information concerning the nature of the interaction may be obtained by energy partitioning techniques. There are several ways of splitting the energy into parts, each of them stressing another aspect. One particular technique has already been described in Chapter II, and closely resembles the partitioned potential methods mentioned in Chapter IV. It permits a decomposition of the total binding energy into an electrostatic, a delocalization, and an exchange term, and is well suited for a discussion of the forces in ion–molecule complexes (see, e. g. Figs. 4 and 5).

Another kind of analysis due to Clementi [86] allows the partition of the interaction energies into contributions from the individual atoms or groups of atoms in a given complex.

Starting from the expression for the total energy [57] of a system containing 2n electrons

$$E = \sum_i^n 2\,h_i + \sum_{i>j}^n \sum^n P_{ij} + E_{\text{NUC}}{}^{\text{p)}} \qquad (47)$$

which is valid for closed shell systems, he rewrites this sum in the form:

$$
\begin{aligned}
E = 2\,\sum_i (\sum_A h_{iA} + \sum_{AB} h_{iAB} + \sum_{ABC} h_{iABC}) + \sum_i \sum_j (\sum_A P_{ijA} + \sum_{AB} P_{ijAB} + \\
+ \sum_{ABC} P_{ijABC} + \sum_{ABCD} P_{ijABCD}) + \sum_{AB} E_{NA,NB}
\end{aligned}
\qquad (48)
$$

The indices A, B, C, and D run over all atoms,

h_{iA}, h_{iAB}, h_{iABC} are the one-, two-, and three-center contributions to h_i.

$E_{NA,NB} = \dfrac{Z_A\,Z_B}{|R_A - R_B|}$ is the contribution of centers A and B to the nuclear repulsion.

P_{ijA}, P_{ijAB}, P_{ijABC}, and P_{ijABCD} are the n-center contributions to the electron interaction integrals P_{ij}, $n = 1, 2, 3, 4$.

Now we can define contributions of atoms (A) or groups $(AB\ldots N)$ to the total energy by collecting those terms of the sum which are labelled exclusively with indices referring to the specific atom A or the group $AB\ldots N$, respectively. Likewise the energy of interaction of two sets of atoms is the sum of those terms which are labelled exclusively with indices referring to both sets simultaneously. Later Kistenmacher et al. [87,198] used this formalism extensively in their studies on alkali ion–water interactions. One of the most interesting conclusions is that at the equilibrium geometries the energy of the cation is changed in the field of the water molecule by approximately equal amounts for all the cations investigated. Provided this trend holds in the full series of alkali ions one is tempted to predict that the gain in polarization energy in Rb^+ and Cs^+ balances the loss of interaction energy due to increased ion size. Thus the energies of interaction for $K^+\cdot OH_2$, $Rb^+\cdot OH_2$, and $Cs^+\cdot OH_2$ may be roughly the same [198].

---

p) P_{ij} is defined as $P_{ij} = (2\,J_{ij} - K_{ij})$ and the h_i, J_{ij}, and K_{ij}'s are integrals over molecular orbitals of the form:

$$h_i = \int \phi_i(r_1)\left(-\frac{1}{2}\nabla_1^2 - \sum_k \frac{Z_k}{|r_1 - R_k|}\right)\phi_i(r_1)\,d\tau_1$$

$$J_{ij} = \int\int \phi_i(r_1)\,\phi_i(r_1)\,\frac{1}{|r_1 - r_2|}\,\phi_j(r_2)\,\phi_j(r_2)\,d\tau_1\,d\tau_2$$

$$K_{ij} = \int\int \phi_i(r_1)\,\phi_j(r_1)\,\frac{1}{|r_1 - r_2|}\,\phi_i(r_2)\,\phi_j(r_2)\,d\tau_1\,d\tau_2$$

E_{NUC} is the nuclear repulsion energy, r_i and R_j here denote the coordinates of the i-th electron and j-th nucleus, respectively.

A similar procedure dividing the energy into single-, two-, three- etc. particle contributions was used by Hankins *et al.* [89] to discuss nonadditivities in small water clusters, Eq. (49, 50).

$$E(\mathbf{r}_1 \ldots \mathbf{r}_N) = \sum_i E^{(1)}(\mathbf{r}_i) + \sum_{ij} V^{(2)}(\mathbf{r}_i, \mathbf{r}_j) + \sum_{ijk} V^{(3)}(\mathbf{r}_i, \mathbf{r}_j, \mathbf{r}_k) + \ldots \qquad (49)$$

There $E^{(1)}$ is simply defined as the energy of the i-th isolated subsystem:

$$E^{(1)}(\mathbf{r}_i) = E(\mathbf{r}_i) \qquad (50\,a)$$

and consequently two- and three-body contributions are obtained by the following expressions:

$$V^{(2)}(\mathbf{r}_i, \mathbf{r}_j) = E(\mathbf{r}_i, \mathbf{r}_j) - E^{(1)}(\mathbf{r}_i) - E^{(1)}(\mathbf{r}_j) \qquad (50\,b)$$

$$V^{(3)}(\mathbf{r}_i, \mathbf{r}_j, \mathbf{r}_k) = E(\mathbf{r}_i, \mathbf{r}_j, \mathbf{r}_k) - \sum_{l=ijk} E^{(1)}(\mathbf{r}_l) - \sum_{l>m=ijk} V^{(2)}(\mathbf{r}_l, \mathbf{r}_m) \qquad (50\,c)$$

The major difference between both techniques concerns the use of constant single particle energies in Eq. (49). Therefore, the contributions of single particle polarization is included in the higher terms. We will use this method in Section V/A2 for bigger clusters.

A very illustrative and useful way to study the rearrangements of electron distributions is represented by charge density difference diagrams. Fig. 11 shows the spatial details of $\Delta\varrho$ (Eq. 34) in the (zy) plane of the Li^+–OH_2 complex [208].

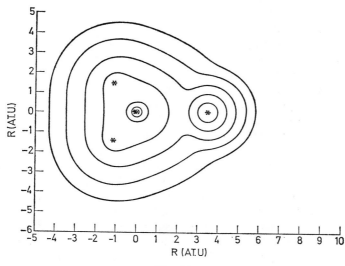

Fig. 11 a

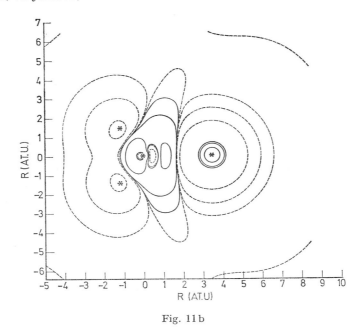

Fig. 11b

Fig. 11a and b. a) Total electron density of the systems H_2O–Li^+ in the plane of the water molecule. b) Electron density differences upon complex formation shown in the same plane (maps from Ref. [208])

We observe a gain in electron density in the lone-pair region of the water molecule and a corresponding loss at the hydrogen nuclei and at the complex periphery. Some charge is found to envelop the Li^+ core without penetrating it. The $Li^+ - 1s$ core region does not seem to be as rigid as one would suggest on physical reasons. As mentioned in Chapter II, the integrated charge density difference function $\Delta\varrho(z)$ [Eq. (35)] is better suited for a rough description of charge migrations in the complex, although it necessarily obscures some information about spatial details. The z-axis is defined as before in Chapter II. Figs. 12 and 13 show $\Delta\varrho(z)$ for the complexes $Li^+ \cdot OH_2$ and $Be^{2+} \cdot OH_2$ in their equilibrium positions.

For singly charged cations nearly no charge is transferred to the ligand (Table 6). In the case of Be^{2+}, however, a considerable effect is observed, supporting the view that nonnegligible covalent contributions are present. This has been argued earlier on the basis of population analysis and comparison with classical models [111]. Figures 14 and 15 show the polarization effects as a function of ion–ligand distance. At large distances only polarization of the ligand occurs, leading to a single density fluctuation. The amplitude of the fluctuation increases almost symmetrically as the ion approaches. At shorter distances, however, the shape of the curve changes rather unexpectedly. Instead of being transferred to the ion, the electrons in the region of overlap are pushed back toward the oxygen of the ligand. At the same time deshielding in the region of the hydrogens increases continuously. A similar effect has already been observed in hydrogen-

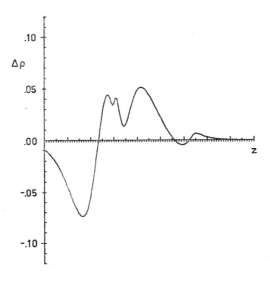

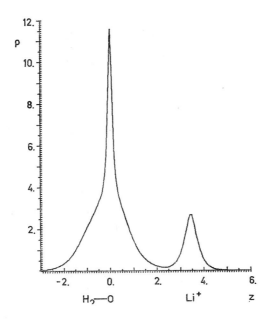

Fig. 12. Integrated electron density and integrated electron density differences in the system H_2O–Li^+. (cf. Ref. [109])

P. Schuster, W. Jakubetz, and W. Marius

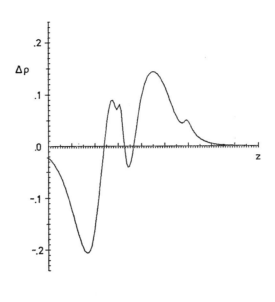

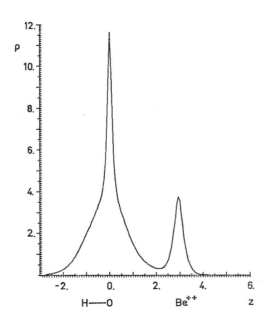

Fig. 13. Integrated electron density and integrated electron density differences in the system $H_2O–Be^{2+}$ (cf. Ref. [111])

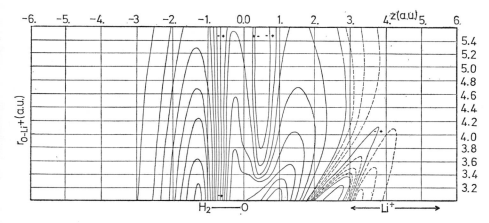

Fig. 14. Integrated electron density differences in the system H_2O-Li^+ as a function of the O–Li^+ distance (cf. Ref. [109]). The spacing of the lines is 0.01 (e_0) for continous lines and 0.0025 (e_0) for broken lines (cf. Ref. [109])

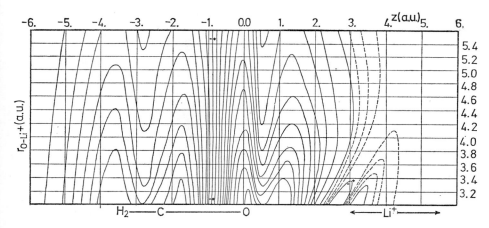

Fig. 15. Integrated electron density differences in the system H_2CO-Li^+ as a function of the O–Li^+ distance (cf. Ref. [109]). The spacing of the lines is 0.01 (e_0) for continuous lines and 0.0025 (e_0) for broken lines

bonded associations [269]. According to the low ionization potentials of the alkali atoms, little energy is gained by charge transfer, and "exchange-forces" due to the inner shell electrons of the alkali cores predominate. For other ions the situation might be quite different.

As Fig. 12 shows, the inner shell electrons of the alkaline ions behave classically like a polarizable spherical charge-density distribution. Therefore it seemed promising to apply a "frozen-core" approximation in this case [194]. In this formalism all those orbitals which are not assumed to undergo larger changes in shape are not involved in the variational procedure. The orthogonality requirement is replaced by a nonlocal pseudopotential. This approximation saved a considerable amount of time and thus the $K^+ \cdot OH_2$ does not require much more computational efforts than the $Li^+ \cdot OH_2$. Moreover, the frozen inner-shell electrons do not appear in the total energies, and less numerical accuracy is necessary. Inevitably the approximations of the model bring about systematic deviations from all-electron SCF calculations. These errors are well understood and can be partly eliminated. Table 11 presents some results obtained by this kind of approach. The assumption of a hard core does not allow the ligands to come close enough to the ion, resulting also in a decrease of binding energy. As expected, the errors increase with increasing atomic number of the metal ion.

b) Other Solvent Molecules. Ammonia is the smallest chemically reasonable unit containing a single lone-pair of electrons. Despite this interesting feature its properties in ionic interactions have been calculated only rarely [199,209,210]. The first *ab initio* calculations on $NH_4^+ \cdot NH_3$ were published by Merlet *et al.* [209] and Delpuech *et al.* [210]. Both groups investigated the potential curve for proton transfer along the NHN axis as well as the dissociation into the subunits. In agreement with available experimental data [127,140] the energy of interaction in $N_2H_7^+$ was found to be somewhat larger than in the corresponding oxygen compound $H_5O_2^+$ (Table 15).

The system $HCN \cdot Li^+$ has been calculated recently [106]. As is shown from both Table 16 and Fig. 18 the responses of both lone pair and π-electrons are well separated. Comparison with Fig. 16 shows that the effects are comparable in magnitude to those occurring in the system $H_2CO \cdot Li^+$. Recent interest in lipophilic macrocyclic compounds, which are well known as ionophores, stimulated a thorough investigation of ion carrier interaction. Structural data suggest ether-, ester-, and carbonyl-oxygens to be the binding sites for the metal ions.

Its moderate size makes the carbonyl group a model system particularly suited for the study of π-systems. Calculations on the $Li^+ \cdot H_2CO$ interaction with medium-sized basis sets have been made. In contrast to $Li^+ \cdot OH_2$, the energy surface shows roughly rotational symmetry [110,113]. In more detail, the mobility of the ion is a little greater in the plane of the molecule than perpendicular to it. Energy partitioning has already been presented in Fig. 5 and demonstrates most of the energy to be of ion-dipolar type. Delocalization of electrons plays a more important role than in the corresponding system with water. As inferred from the integrated density difference curves (cf. Figs. 15 and 16) this is mainly due to internal polarization of the formaldehyde. Some charge transfer to the Li^+ cation also takes place but still remains small in magnitude. An increased polarizability of π-electrons has been demonstrated by calculating the shift of the centers of charge of individual localized MO's [113] (see Table 16). Recently the interaction of formaldehyde with the alkali ions Li^+ through K^+ was investigated using the pseudopotential formalism mentioned above [194]. The restrictions in variational freedom of the wave function are evidently reflected in the deviations from comparable all-electron calculations (see Table 15) but remain small enough for the purpose of comparison.

Functional groups attached to the carbonyl group naturally alter the shape of the energy surface to some extent. Rode and Preuss [212] studied the interaction of Li^+ with formamide. As a result of the amine substituent the mobility

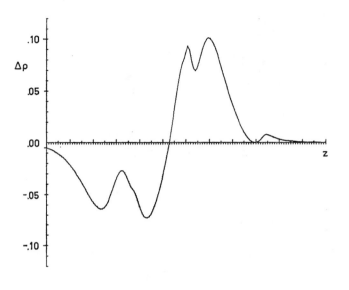

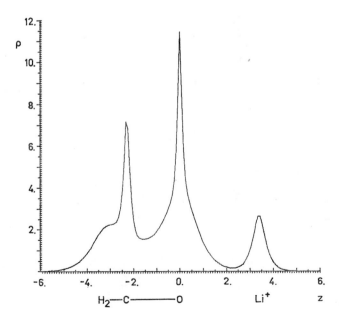

Fig. 16. Integrated electron density and integrated electron density differences in the system H_2CO-Li^+ near the equilibrium geometry (cf. Ref. [109])

Table 15. Cation–solvent complexes with solvent molecules other than water

Ion	Solvent	Ref.	Basis set[1] uncontracted	contracted	Equilibrium geometry[2]	Interaction energy, $-\Delta E$	Remarks
NH_4^+	NH_3	199)	STO3G	—	$R_{NH} = 2.50$ 2.75 2.80	42.2 14.3 16.1	Linear H-bond NH_3 lone pair approaching C_{3v} axis of NH_4^+ NH_3 lone pair bisecting HNH angle of NH_4^+
NH_4^+	NH_3	209)	GTO: N(4/2) H(1) bridge H(1/1)	[4/2] [1] [1/1]	$R_{NN} = 2.71$	36.0	Linear H-bond
NH_4^+	NH_3	210)	GTO: N(8/4/1) H(3/1)	[8/4/1] [3/1]	$R_{NN} = 2.78$	27.6	Linear H-bond
NH_4^+	NH_3	140)	—	—	—	24.8	Exp. value
NH_4^+	$4(NH_3)$	199)	STO3G	—	$R_{NH} = 2.70$	113.6	
NH_4^+	$4(NH_3)$	140)	—	—	—	68.6	Exp. value
Li^+	H_2CO	113)	GTO: C/O(5/3) H(2) Li(4)	[3/2] [2] [3]	$R_{OLi^+} = 1.80$	44.9	CO..Li^+ coaxial
Li^+	H_2CO	194)	GTO: C/O(5/3) H(2) Li(4)	[3/2] [2] [3]	$R_{OLi^+} = 1.84$	43.0	Pseudopotential calculation
Li^+	H_2CO	110)	GTO: C/O(9/5) H(4) Li(9/1)	[5/3] [3] [5/1]	$R_{OLi^+} = 1.77$	43.2	
Li^+	H_2CO	109)	GTO: C/O(9/5/1); H(4) Li(9/1)	[5/3/1] [3] [5/1]	$R_{OLi^+} = 1.78$	44.2	

Table 15. (continued)

Ion	Molecule	Ref	GTO basis	Contraction	R	Energy	Description
Na+	H_2CO	194)	GTO: C/O(5/3) H(2) Na+(6)	[3/2] [2] [3]	$R_{ONa^+} = 2.29$	29.7	Pseudopotential calculation
K+	H_2CO	194)	GTO: C/O(5/3) H(2) K(7)	[3/2] [2] [4]	$R_{OK^+} = 2.82$	20.4	Pseudopotential calculation
Li+	$HCONH_2$	212)	GTO: N/C/O(4/2) Li(4/1)	[4/2] [4/1]	$R_{OLi^+} = 1.71$	55.9	Li+ colinear with CO bond
Li+	$2(HCONH_2)$	213)	GTO: N/C/O(4/2) Li(4/1)	[4/2] [4/1]	$R_{OLi^+} = 1.83$ $R_{NLi^+} = 2.19$	103.0	Li+ as inversion center of the chelate complex
Li+	HCN	106)	GTO: C/N(7/3/1) H(3/1) Li(6/1)	[4/2/1] [2/1] [3/1]	$R_{NLi^+} = 1.94$	39.3	linear geometry
Li+	HCN	106)	GTO: C/N(9/5/1) H(4/1) Li(6/1)	[5/3/1] [3/1] [3/1]	$R_{NLi^+} = 1.96$	37.0	Linear geometry
Na+	CH_3COCH_3	197)	STO3G	–	$R_{ONa^+} = 1.99$	38.8	35° deviation from CO direction in the ester plane
Na+	$CN_3NHCOCH_3$	197)	STO3G	–	$R_{ONa^+} = 1.95$	49.7	35° deviation from CO direction in the amide plane
K+	CH_3OCOCH_3	197)	STO3G	–	$R_{OK^+} = 2.40$	25.5	35° deviation from CO direction in the ester plane
K+	$CH_3NHCOCH_3$	197)	STO3G	–	$R_{OK^+} = 2.35$	35.1	35° deviation from CO direction in the amide plan

1) For the notation of the basis set see Table 11.
2) All lengths are in (Å), energies in (kcal/mole).

Table 16. Shift of the centers of charge of the localized orbitals upon complex formation[1]

Orbital	Complex		
	H$_2$O–Li$^+$	H$_2$CO–Li$^+$	HCN–Li$^+$
1sLi$^+$	0.0021	0.0006	0.0035
1sO	0.0002	0.0001	—
1sN	—	—	0.0003
1sC	—	0.0001	0.0002
σ CO	—	0.0073	—
σ C–N	—	—	0.0181
σ C–H	—	0.0216	0.0403
σ O–H	0.0451	—	—
Lone pair O	0.0221	0.0262	—
Lone pair N	—	—	0.0884
π Electrons	—	0.1162	0.1200
Reference	224)	224)	106)

[1] Shifts in [Å] in the direction to the Li$^+$ cation.

of the Li$^+$ ion in the plane of the molecule is much more pronounced than in the case of formaldehyde. The authors report two energy minima, one corresponding approximately to a carbonyl- and the other corresponding to an amide binding site. The stable complex is represented by the former structure. Pullman [197] investigated in detail the interactions of N-methylacetamide and of N-methylacetate with sodium and potassium ions. Amide carbonyl seems to be preferred over ester carbonyl. This result clearly reflects the decrease in the constant of complex formation upon replacing NH-groups by ester oxygens, e.g. in the nonactine or valinomycine molecule [115,214]. The affinities of alkaliy ions are somewhat lower for ester carbonyl than for water. As in the case of formamide, enhanced mobility of the ion is found. Here the equilibrium position of the ion is shifted from the C—O axis towards the α-methyl group. Generally, non-planar complexes are found to be appreciably less stable than planar ones.

2. Interaction of Cations with Two or More Solvent Molecules

For an adequate description of solvated ions two-body potentials are not sufficient in most cases. It has been shown from calculations on water networks that the postulate of pairwise additivity of the interaction energy in higher aggregates is often strongly violated by "cooperative" effects [89,216]. Therefore a precise knowledge of three- and even more-body potentials would be desirable. Unfortunately not only the computing time necessary for the calculation of a single point on the energy surface increases drastically, but also the number of degrees of freedom of the system is enlarged. Nevertheless some examples have been studied even with extended basis sets: Li$^+$(H$_2$O)$_n$, $n = 1$ to 6 [90,162,217], Na$^+$(H$_2$O)$_n$, $n = 1$ to 4 [162], K$^+$(H$_2$O)$_n$, $n = 1$ to 3 [162] and Be^{2+}(H$_2$O)$_n$, $n = 1$ to 3 [90]. The results are summarized in Table 17. In the Li$^+$–water complexes two main

Table 17. Interactions of cations with several water molecules

Ion	Number of H_2O molecules in the complex	Basis set[1]	Ref.	Geometry[2]	Stabilization energy[2] $-\Delta E$
Li^+	2	O(11/7/4) [5/4/1] H(6/1) [3/1] Li(11/2) [5/2]	217)	See Fig. 19A $R_{OLi^+} = 1.92$	67.54
				See Fig. 19B $R_{OLi^+} = 1.85$ $R_{OO} = 2.70$	52.4
Li^+	2	O(9/5/1) [4/2/1] H(4/1) [2/1] Li(9/3) [4/2]	90)	See Fig. 19B $R_{OLi^+} = 1.85$ (not varied) $R_{OO} = 2.70$	52.4
Li^+	2	O(11/7/1) [5/4/1] H(6/1) [3/1] Li(11/2) [4/2]	162)	See Fig. 19A $R_{OLi^+} = 1.87$	64.8[3]
Li^+	2	—	130)	—	59.8 exp. value
Li^+	3	See two rows above	162)	Trigonal $R_{OLi^+} = 1.90$	88.7[3]
Li^+	3	—	130)	—	80.5 exp. value
Li^+	4	See four rows above	162)	Tetrahedral $R_{OLi^+} = 1.94$	105.9[3]
Li^+	4	—	130)	—	96.9 exp. value
Li^+	5	See six rows above	162)	Trigonal bipyramid $R_{OLi^+} = 2.02$	112.7[3]
Li^+	5	—	130)	—	110.8 exp. value
Li^+	6	See eight rows above	162)	Octahedral $R_{OLi^+} = 2.08$	124.6[3]
Li^+	6	—	130)	—	122.9 exp. value
Be^{2+}	2	O(9/5) [4/2] H(4) [2] Be^{2+}(9/3) [4/2]	90)	Linear, planar $R_{OBe^{2+}} = 1.50$ (not varied)	292.9
Be^{2+}	3	Same as above	90)	Two H_2O in the xy plane pointing along x and $-x$ axis, third H_2O in the xz plane pointing along z axis. $R_{OBe^{2+}} = 1.50$ not varied	303.9

Table **17** (continued)

Ion	Number of H_2O molecules in the complex	Basis set[1])	Ref.	Geometry[2])	Stabilization energy[2]) $-\Delta E$
Na⁺	2	O(11/7/1) [5/4/1] H(6/1) [3/1] Na⁺(13/8/1) [7/4/1]	162)	See Fig. 19A $R_{ONa^+} = 2.23$	47.9[3])
Na⁺	2	—	130)	—	43.8 exp. value
Na⁺	3	Same as two rows above	162)	Trigonal $R_{ONa^+} = 2.25$	66.8[3])
Na⁺	3	—	130)	—	59.6 exp. value
Na⁺	4	See four rows above	162)	Tetrahedral $R_{ONa^+} = 2.28$	82.0[3])
Na⁺	4	—	130)	—	73.4 exp. value
K⁺	2	O(11/7/1) [5/4/1] H(6/1) [3/1] K(17/11/1) [11/7/1]	162)	See Fig. 19A $R_{OK^+} = 2.69$	33.5[3])
K⁺	2	—	130)	—	34.0 exp. value
K⁺	3	See two rows above	162)	Trigonal $R_{OK^+} = 2.69$	47.5[3])
K⁺	3	—	130)	—	47.2 exp. value
NH_4^+	2	STO3G	199)	Linear H-bridge $R_{NO} = 2.45$	56.8
NH_4^+	2	—	140)	—	32.0 exp. value
NH_4^+	3	STO3G	199)	Linear H-bridge $R_{NO} = 2.55$	66.3
NH_4^+	3	—	140)	—	45.4 exp. value
$NH_4^\pm$	4	STO3G	199)	Linear H-bridge $R_{NO} = 2.60$	70.4
NH_4^+	4	—	140)	—	57.6 exp. value
$(Met)_4N^+$	2	STO–3G	200)	bisecting CNC angle Axial along CN bond	12.5 19.2

Table 17 (continued)

Ion	Number of H_2O molecules in the complex	Basis set[1]	Ref.	Geometry[2]	Stabilization energy[2] $-\Delta E$
(Met)$_4$N$^+$	3	STO3G	200)	Bisecting CNC angle	18.2
				Axial	27.6
(Met)$_4$N$^+$	4	STO3G	200)	Bisecting CNC angle	23.5
				Axial	35.3
(Met)$_3$N$^+$ Et	3	STO3G	200)	—	26.7
(Met)$_2$N$^+$ (Et)$_2$	2	STO3G	200)	$R_{NO} = 3.6$ 30° above and below N(Met)$_2$ plane	16.0

[1] For the notation of the basis set see Table 11.
[2] All lengths in [Å], energies in (kcal/mole).
[3] The geometries were obtained from Monte Carlo calculations. (cf. Ref. [162]) where the H_2O molecules are constrained to have all equal ion–oxygen distance. The ratio of the multibody effects is for Li$^+$(H$_2$O)$_n$ 1./3.3/6.5/10.7 for Na$^+$(H$_2$O)$_n$ 1.0/2.9/5.2 and for K$^+$(H$_2$O)$_n$ 1.0/2.4.

structures are of special interest (cf. Fig. 19): (1) structure A with C_{2v}-symmetry, where the lone pairs of the oxygen point toward the ion, and (2) structure B, where the Li$^+$ is linked to the donor site of the water dimer.

The first structure which turns out to be the more stable one shows a smaller incremental binding energy and a larger Li–O distance than was found in Li$^+ \cdot$ OH$_2$. This has to be expected since the water–water dipolar interaction is repulsive and polarization of the Li$^+$ ion is reduced.

Density difference diagrams shown in Fig. 17 indicate that polarization of the water remains almost unaffected. In the second geometry investigated (B), we find both ion–molecule interaction and hydrogen bonding. In the field of the Li$^+$, the acidity of the protons in the water molecule adjacent to the cation is substantially increased. We expect therefore the hydrogen bond to the second water molecule to be stronger here than in the isolated water dimer. Using the partitioning method of Hankins et al. [89], Kollman and Kuntz [90] arrived at a hydrogen bond energy of 8.7 kcal/mole compared to 5 kcal/mole in the water dimer [211]. The energy of interaction between Li$^+$ and the neighboring water molecule remains unaffected by the presence of a second water molecule.

If two-body potentials and the three-body contribution of Li$^+$(H$_2$O)$_2$ are taken into account the optimum coordination number for a static Li$^+$(H$_2$O)$_n$ complex turns out to be 4. For the most stable conformation of Li$^+$(H$_2$O)$_6$ they found that two water molecules are bound in a second, outer hydration layer.

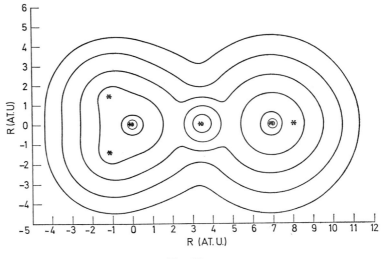

Fig. 17 a

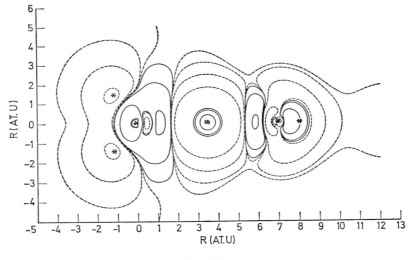

Fig. 17 b

Fig. 17a and b. a) Total electron density of the system $H_2O–Li^+–OH_2$ in the plane of one water molecule. b) Electron density differences upon complex formation as shown in the same plane (maps from Ref. [208])

Although experimental evidence seems to support these results, the fact that the optimum coordination number obtained depended to a large extent on the use of three body contributions warns against uncritical belief in the convergency of the expansion [Eq. (49)] and stresses the preliminary nature of the conclusions. When two-body potentials were considered exclusively, Kollman and Kuntz [219]

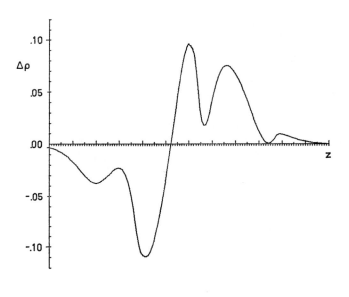

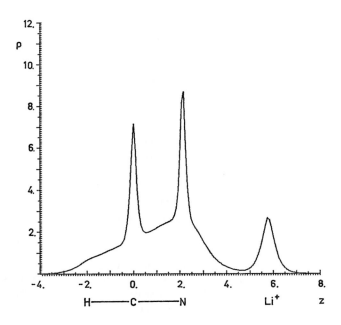

Fig. 18. Integrated electron density and integrated electron density differences in the system HCN–Li+ near the equilibrium geometry (cf. Ref. [106]).

found an optimum coordination number of six for $Li^+(H_2O)_n$. The importance of three-body contributions at some configurations of $M^+(H_2O)_2$ complexes was also pointed out by Kistenmacher et al. [162]. They determined stable arrangements for $M^+(H_2O)_n$ at room temperature by a Monte-Carlo procedure. Out of all three-body contributions, the potential of $Li^+(H_2O)_2$ is most important (85% of the three-center energy) and amounts to 9% of the total stabilization energy. This justifies a posteriori the neglect of $(H_2O)_3$ terms by Kollman and Kuntz [219]. Furthermore, Kistenmacher et al. [162] estimated relative magnitudes of many body effects in complexes $M^+(H_2O)_n$, as shown in Table 17. One might believe to find stronger effects in the case of multiply charged ions. Fortunately, the three-body term $(H_2O)_3$, and the four-body contribution $Be^{2+}(H_2O)_3$ turned out to be comparatively small [90].

As mentioned above, Kistenmacher et al. [162] investigated the structure of solvent shells on a statistical mechanical basis. Adopting the assumption of pairwise additivity of intermolecular forces they studied small clusters of ions with up to ten water molecules. At absolute zero temperature they obtained maxima of stabilization energies with a number of 4,5–6 and 5–7 water molecules in the first solvation layers for the Li^+, Na^+, and K^+ ions, respectively. Additional solvent molecules are more likely to form a second solvation shell than to interact directly with the ion. For tetracoordinated complexes of Li^+, Na^+, and K^+ room temperature was simulated by a Monte Carlo procedure. At this temperature, configurations with somewhat higher energy than the minimum values are populated significantly and therefore the resulting macroscopic stabilization enthalpy is substantially smaller in absolute value. Exchange of water molecules between the first and second hydration layers was found to occur readily and in a cooperative way, in contrast to the results that we would expect for rigid, non-dynamical complex geometries. In the case of $Li^+(H_2O)_n$ an exchange of one water molecule between the first and second layer would require an activation energy of about 28 kcal/mole at a frozen geometry of the rest of the complex [162].

A molecular dynamics study of Heinzinger and Vogel [183] uses an empirical effective pair potential. It is mentioned here for reasons of comparison. They solved the equations of motion for an assembly of 9 LiCl molecules interacting with 198 water molecules at a temperature of 272 °K. Their assumptions lead to a coordination number of 5.5 for Li^+ and 6.0 for Cl^- which is in fair agreement with X-ray scattering values 4 ± 1 and 6 ± 1, respectively [175]. Due to the high concentration choosen, no evidence is found for a second solvation layer. A significant change towards higher disorder for the $(H_2O)_2$ correlation function, relative to pure water at the same temperature, is reported. As an example of a multiply solvated polyatomic cation Pullman and Armbruster [199] dealt with the hydration and ammoniation of the ammonium ion. The results are found in Table 17 and show the expected behavior. The influence of alkyl substitution is reasonably reflected. The resulting binding energies and bond lengths are lowered and increased respectively compared to the unsubstituted ammonium ion. Interestingly, ammonia is the preferred solvent molecule only in the first solvation layer. Water dominates in subsequent shells. Similar results were obtained by recent experimental investigations [140]. As indicated, the ion-formaldehyde interaction was investigated as a first step toward an understanding of ion-binding to ionophores.

In some of the biologically important ion carriers, however, the ion is surrounded by several carbonyl groups. Strongly simplified model calculations were made by Pullman and Schuster [110]. Assuming pairwise additivity in the first solvation shell they tried to approximate the binding site of the ionophore by an octahedral arrangement of six formaldehyde molecules. Multiple minima of the energy hypersurface occur inside the cage provided the exchange contribution to the interaction energy is negligible at the LiO distance assumed. This is true, however, only for very large cages. At smaller distances similar to those actually observed in most ionophores, the exchange contribution dominates and the energy surface for Li-motion shows only a single minimum. Considering the structural data for nonactine metal–ion complexes [218] the assumption of a rigid macrocyclic ligand does not seem to be fulfilled in most cases. Therefore one has to use a more sophisticated model in order to give a correct description of ion-binding to carrier molecules.

The interaction of two formamide molecules with Li^+ has been studied by Rode [213] using a small GTO basis set. He considered two types of geometries in both 1:1 and 1:2 complexes. In the most stable structure of the 1:1 complex the Li^+ cation is situated on the CO connection line. In a "chelate-like" structure the Li^+ ion occupies a position which is almost equidistant from the O and N atoms. Although this type of 1:2 complex is found to be 70 kcal/mole less stable than the other structure with a linear OLiO arrangement it might be important for larger central ions.

3. Results Obtained with Semiempirical Procedures

At is does not seem likely that accurate all-electron calculations on higher aggregates will become available in the near future, recourse has to be made to semiempirical procedures. The CNDO-, INDO- and the extended Hückel methods (EHM) have been used most widely. The assumptions made in these methods may sometimes lead to unrealistic results, like strange geometries and grossly overemphasized charge transfer [220–223], see also Table 18. Within the framework of CNDO, the most stable structure of $Li^+ \cdot OH_2$ is found to be pyramidal with the Li^+ placed more closely to the H atoms than to O [222]. With $Li^+ \cdot OCH_2$ a similar disappointing CNDO/2-equilibrium geometry was obtained [224]. Nevertheless careful use might permit discussion of trends to be expected in a series of complexes and in higher aggregates. Therefore we also present some results from these approaches.

Work analogous to that discussed above in Sections 1 and 2 has been performed. Most of these efforts aimed for an estimation of solvation energies and a determination of preferred geometries in solvent cages. Table 18 summarizes CNDO/2-results derived from simple aggregates where also nonempirical results are available. Fairly correct energies of interactions can be obtained by the semiempirical method provided some constraints are replaced upon geometrical variations, e. g. planarity of the complexes must be imposed. Comparison of Tables 11 and 18, and density difference curves (Figs. 12, 16, 20) shows the existence of a systematically too large charge transfer. As the alkali ions are represented merely as screened nuclei in the CNDO formalism, no polarization of inner shell electrons is possible.

Table 18. Semiempirical (CNDO/II) studies on cation–water interaction

Ion	Number of molecules	Ref.	Charge transferred to the ion	Cage geometry[1]) R_{OX^+}	Interaction energy[2]) $-\Delta E$
Li+	1	220)	−0.160	Planar C_{2v} 2.36	45.0
Li+	1	223)	−0.163	Planar C_{2v} 2.35	45.7
Li+	1	222)	−0.192	116° deviation from HOH bisectrix C_{2v} pyramidal 2.42	51.6
Li+	2	220)	−0.307	Planar C_{2v} 2.37	86.9
Li+	2	223)	−0.312	Planar C_{2v} 2.37	87.9
Li+	2	222)	−0.388	121° deviation from each HOH angle bisectrix, Li+ as inversion center 2.43	102.9
Li+	3	220)	−0.439	Trigonal C_{3v} 2.39	127.0
Li+	4	221,223)	−0.556	Tetrahedral 2.41	164.3
Li+	4	221)	−0.556	Tetrahedral 2.41	163.0
Li+	6	223)	−0.689	Octahedral 2.48	217.7
Li+	6	221)	−0.895	Octahedral 2.49	216.0
Li+	8	223)	−0.755	Archimed. antiprisma 2.55	257.0
Na+	1	220)	−0.075	Planar C_{2v} 3.11	20.6[2])
Na+	1	223)	−0.173	Planar C_{2v} 2.92	32.9
Na+	1	221)	−0.126	96° deviation from HOH bisectrix, pyramidal 2.98	34.9
Na+	2	220)	−0.144	Planar C_{2v} 3.13	40.25[2])
Na+	2	223)	−0.224	Planar C_{2v} 2.92	64.7
Na+	2	221)	−0.252	102° deviation from each HOH bisectrix Na+ as inversion center 3.0	69.7
Na+	3	220)	−0.214	Trigonal C_{3v} 3.13	59.5[2])
Na+	4	221,223)	−0.436	Tetrahedral 2.93	125.9
Na+	4	220)	−0.260	Tetrahedral 3.14	77.7[2])
Na+	6	223)	−0.619	Octahedral 2.95	181.1
Na+	6	221)	−0.487	Octahedral 3.19	109.0[2])
Na+	8	223)	−0.747	Archimed. antiprisma 2.98	229.6
Be2+	1	223)	−0.524	lin C_{2v} 1.70	217.6
Be2+	2	223)	−0.993	lin C_{2v} 1.72	411.8
Be2+	4	223)	−1.438	Tetrahedral 1.80	682.9
Be2+	4	221)	−1.420	Tetrahedral 1.80	678.0
Be2+	6	223)	−1.504	Octahedral 1.89	821.4
Be2+	6	221)	−1.89	Octahedral 2.0	818.0
Be2+	8	223)	−1.471	Archimed. antiprisma 1.99	891.2
Mg2+	4	221)	−0.636	Tetrahedral 2.61	202.0[2])
Mg2+	6	221)	−1.10	Octahedral 2.67	270.0[2])

Table 18 (continued)

Ion	Number of molecules	Ref.	Charge transferred to the ion	Cage geometry[1] R_{OX+}	Interaction energy[2] $-\Delta E$
NH_4^+	1	[220]	−0.093	Linear H-bond 1.38	25.7
NH_4^+	2	[220]	−0.152	Linear H-bonds 1.42	47.0
NH_4^+	3	[220]	−0.191	Linear H-bonds 1.44	65.1
NH_4^+	4	[220]	−0.219	Linear H-bonds 1.46	81.0
NH_4^+	4	[221]	—	Linear H-bonds 1.44	89.4
$CH_3NH_3^+$	1	[220]	−0.078	Linear H-bond 1.41	21.61
$CH_3NH_3^+$	2	[220]	−0.130	Linear H-bonds 1.44	40.25
$CH_3NH_3^+$	3	[220]	−0.170	Linear H-bonds 1.46	56.5
$CH_3NH_3^+$	3	[221]	—	Linear H-bonds 1.44	62.0
$(CH_3)_2NH_2^+$	1	[220]	−0.066	Linear H-bonds 1.44	18.8
$(CH_3)_2NH_2^+$	1	[220]	−0.115	Linear H-bonds 1.46	35.4

[1] All lengths in (Å), energies in (kcal/mole).
[2] A modified CNDO procedure was used.

Several models of solvent cages have been proposed [221–223]. Burton and Daly [221] were the first to investigate tetrahedrally and octahedrally coordinated ions. Their results are summarized in Table 18. The greater apparent size of octahedrally hydrated ions is also observed in the more accurate calculations and might be due to dipole–dipole and steric interactions among the ligands. Force constants, which could give an idea of the stability of the cage are larger in octahedrally than in tetrahedrally coordinated ions. Obviously force constants increase with increasing ionic charge [221]. Lischka et al. [222] and Russegger et al. [223] investigated cation–water interactions for 1, 2, 4, 6, and 8 water molecules in the first layer and examined the bonding properties of water molecules bound in outer shells. Again incremental binding energies decrease and bond lengths increase with increasing number of the ligands. On the other hand, the total interaction energy grows monotonically with the number of ligands up to a coordination number of eight.

The effect of the ion on the strength of the hydrogen bond between water molecules dies off rapidly in the outer shells of the clusters. Almost the same values are observed in the third solvation layer and in tetrahedral water clusters. Chain structures discussed by Burton and Daly [220] show an analogous behavior (Table 19).

EHM has also been applied for the calculations of optimum coordination numbers [226] in hydrated alkali ions (Table 20). The affinities of alkali ions to NN'-dimethylacetamide and methylacetate were estimated by Kostetsky et al. [227], using the CNDO/2 procedure (Table 21). As long as the ions are constrained to lie in the peptide- or ester planes, correct trends are obtained, but relaxation of this constraint reveals serious discrepancies to *ab initio* calculations. The CNDO procedure artificially stabilizes structures with nonpolar bonding geometries,

Table 19. Chain structures for cation–water complexes calculated in the CNDO/II approximation

Number of H_2O molecules[1]	1		2			3				4				
Ion	$R_{O_1C^+}$	$-\Delta E$	$R_{O_1C^+}$	$R_{H_1O_2}$	$-\Delta E$	$R_{O_1C^+}$	$R_{H_1O_2}$	$R_{H_2O_3}$	$-\Delta E$	$R_{O_1C^+}$	$R_{H_1O_2}$	$R_{H_2O_3}$	$R_{H_3O_4}$	$-\Delta E$
Li^+	2.36	45.0	2.30	1.41	64.6	2.28	1.38	1.45	78.7	2.28	1.37	1.42	1.47	90.7
Na^+	3.11	20.6	3.03	1.46	35.1	3.01	1.43	1.47	47.4	3.00	1.42	1.44	1.48	58.6
NH_4^+	1.38	25.6	1.34[2]	1.42	43.0	1.33	1.38	1.46	56.6					

[1] All lengths in (Å), energies in (kcal/mole). The structures are all assumed planar and linear H-bonded, each H_2O molecules acts as proton donor for the following. All data are from Ref. 220).

[2] R_{HH_1}.

Table 20. Extended Hückel calculations on cation–water complexes[1])

Ion	Number of water molecules	Cage geometry[2])	Total stabilization energy $-\Delta E$	Remarks
Li$^+$	4	Tetrahedral $R_{OLi^+} = 1.44$	140.0	
Li$^+$	6	Octahedral $R_{OLi^+} = 1.67$	185.0	
Li$^+$	8	Cubic $R_{OLi^+} = 2.03$	145.0	
Na$^+$	4	Tetrahedral $R_{ON_a^+} = 1.62$	152.0	Including d-type functions on Na$^+$
Na$^+$	6	Octahedral $R_{ON_a^+} = 1.76$	230.0	
Na$^+$	8	Cubic $R_{ON_a^+} = 2.05$	215.0	
K$^+$	4	Tetrahedral $R_{OK^+} = 2.68$	85.0	Including d-type functions on K$^+$
K$^+$	6	Octahedral $R_{OK^+} = 2.63$	128.0	
K$^+$	8	Cubic $R_{OK^+} = 2.56$	121.0	

[1]) All values from Ref. [226]).
[2]) Lengths in [Å], energies in [kcal/mole].

similar to those discussed above for smaller ligands like H_2O and H_2CO. Using mainly semiempirical techniques, Rode investigated ion solvation in formic acid [229–231]. Again these CNDO/2 calculations lead to an equilibrium geometry of the complex, which deviates largely from that obtained by *ab initio* calculations performed later on [232]). Since a small basis set was applied, a definite conclusion on the actually preferred structure cannot yet be drawn.

B. Anion-Solvent Interactions

1. Anion Hydration Studies

Although interesting, the anion–molecule interactions have not been treated as extensively by *ab initio* methods as cation–molecule interactions. This may be partially due to convergence difficulties which frequently arise in H. F. calculations including negatively charged species. The only anions discussed in some detail are

Table 21. CNDO/II calculations on cation–solvent interaction with solvent molecules other than water

Ion	Solvent	Ref.	Equilibrium[1] geometry	Interaction[2] energy, $-\Delta E$	Remarks
Li⁺	H_2CO	224) 110)	$R_{OLi^+} = 2.32$	51.5	C=O—Li⁺ coaxial
Li⁺	CH_3OCOCH_3	214)	$R_{OLi^+} = 2.10$	78.0	
Li⁺	$(CH_3)_2NCOCH_3$	214)	$R_{OLi^+} = 2.10$	83.0	X⁺ constrained to move along CO axis. (The absolute
Na⁺	CH_3OCOCH_3	214)	$R_{ONa^+} = 3.00$	60.0	minimum deviates from the ester resp. amide plane).
Na⁺	$(CH_3)_2NCOCH_3$	214)	$R_{ONa^+} = 3.00$	63.0	
Li⁺	HCOOH	229,230)	$R_{X^+C} = 2.41$	103.0	
Li⁺	2(HCOOH)	229,230)	2.41	202.2	
Li⁺	4(HCOOH)	229,230)	2.57	276.0	Cation approaching along bisectrix of $\underset{O}{\overset{H}{\diagdown}}\hspace{-1mm} C$ angle.
Na⁺	HCOOH	230)	3.05	70.5	
Na⁺	2(HCOOH)	230)	3.04	140.6	
Na⁺	4(HCOOH)	230)	3.05	276.0	

[1] All lengths in (Å).
[2] Energies in (kcal/mole).

fluoride and chloride, in addition to hydration studies on OH⁻, which are discussed in detail elsewhere [233].

a) One to One Complexes. Equilibrium geometries and energies of interaction
for clusters $X^- \cdot H_2O$ are summarized in Table 22. The question whether hydrogen
bonding (Fig. 19, C, $\delta = 0°$) or ion–dipole (Fig. 19, C, $\delta = 52.3°$) interaction

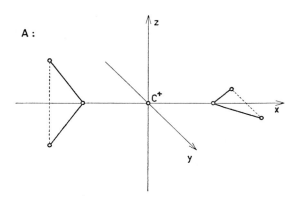

Fig. 19 a

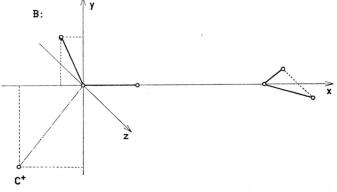

Fig. 19 b

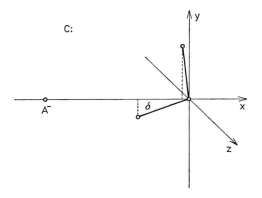

Fig. 19 c

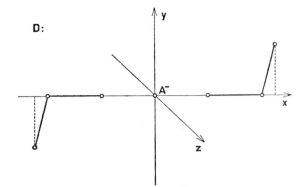

Fig. 19 d

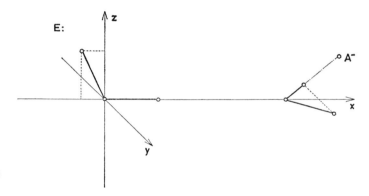

Fig. 19 e

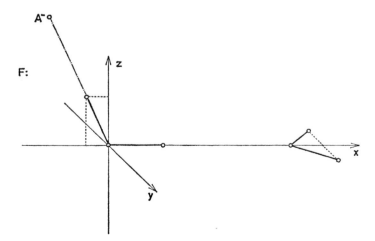

Fig. 19 f

Fig. 19. Geometries for some ion–water systems

Table 22. *Ab initio* calculations on anion–H₂O complexes

Anion	Basis set[2]	Ref.	Equilibrium geometry[1]	Interaction[1] energy, $-\Delta E$
F⁻	STO-4G	195)	Linear H-bond R_{OF^-} = 2.3 (not varied)	29.31
F⁻	GTO: F⁻/O(11/7/1) [5/4/1] H (6/1) [3/1]	203)	Linear H-bond R_{OF^-} = 2.52	24.70
F⁻	GTO: F⁻ (13/8/1) [7/3/1] O (11/7/1) [4/3/1] H (6/1) [2/1]	204)	Bifurcated R_{OF^-} = 2.64 / See Fig. 19 C R_{OF^-} = 2.56 $\delta = 4.5°$	17.24 / 22.60
F⁻		131)		23.3 exp. v
Cl⁻	GTO: Cl⁻ (17/10/1) [9/6/1] O (11/7/1) [4/3/1] H (6/1) [2/1]	204)	See Fig. 19 C R_{OCl^-} = 3.30 $\delta = 14.6°$	11.93
Cl⁻	double ξ STO	205)	See Fig. 19 C R_{OCl^-} = 3.04 $\delta = 0.0°$	19.00
Cl⁻		131)		13.10 exp. v

1) All lengths in (Å), energies in (kcal/mole).
2) For the notation of the basis set see Table 11.

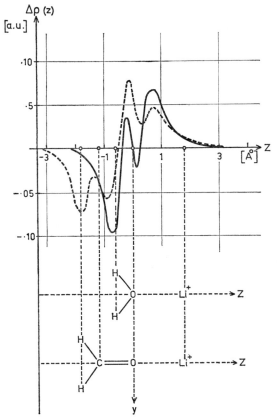

Fig. 20. Integrated electron density differences in the systems H_2O–Li^+ and H_2CO–Li^+ calculated in the CNDO/2 approximation. (Wave functions taken from Ref. [224])

dominates at the equilibrium geometries has been frequently discussed in the past [38,131]. All quantum chemical calculations performed so far confirm the existence of a strong hydrogen bond in $F^-\cdot HOH$, which leads to an almost linear arrangement of the three atoms FHO and a rather high barrier for the in-plane rotation of the water molecule (10 kcal/mole) [204]. The same orientation of the ligand was observed experimentally in solution [172], cf. Chapter III. In the case of $Cl^-\cdot HOH$ the deviation from linearity ($\delta > 0°$) is much larger at the equilibrium geometry. The most striking difference, however, concerns the flexibility of the water molecule in the complex. A barrier of about 1.5 kcal/mole was calculated for in-plane rotation [204]. This result is not too different from that suggested by the electrostatic potential of the water molecule (Fig. 9). For larger ions, like Br^- or I^-, one would expect therefore the classical ion–dipole structure to be preferred. At any rate, within the limit of great distances the classical structures will predominate in all complexes $X^-\cdot H_2O$. The shape of the energy surface and the energy barrier separating the equivalent minima at small distances can be visualized best in Fig. 21.

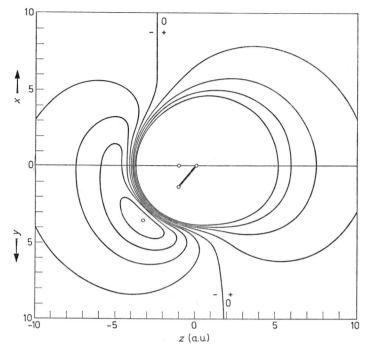

Fig. 21a

Correlation energy and nuclear motion play a similar role as in the corresponding cation complexes: their influence on total binding energies and equilibrium geometries is very small. Relaxation of the ligand geometry has a somewhat larger influence on the energy of interaction, but as can be seen from Tables 13 and 14 the effect still remains small in magnitude compared to the binding energy.

Bond-energy decomposition performed by Kistenmacher et al.[204] suggests a mechanism for the binding in the hydrogen-bonded structure which is determined largely by interaction terms but not by polarization. However, the reverse is true for the structure with C_{2v} symmetry. Figure 22 shows the mutual polarization of both ligand and anion in $F^- \cdot HOH$. Charge is substracted from the bridge proton as it occurs typically in hydrogen-bonded complexes. In addition to a scaling factor, the integrated density difference curve shown in Fig. 23 shows a striking similarity to the curve obtained for the water dimer, cf. Fig. 24. Charge transfer from the anion to the water molecule remains small, although it is greater here by one order of magnitude than in case of $Li^+ \cdot OH_2$, cf. Table 6.

b) *Interactions of Anions with Two or More Solvent Molecules.* Three structures are of particular interest in the configurational space of an anion with two water molecules (Fig. 19, D–F [161]). The first one, denoted with D shows C_{2v} symmetry. The water molecules lie in a common plane, each forming a linear hydrogen bond to the fluoride anion at the inversion center of the complex. In both structures E and

83

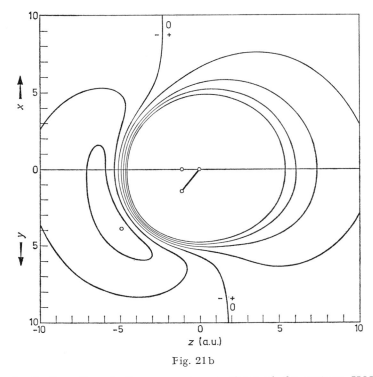

Fig. 21 b

Fig. 21 a–b. Sections through the energy hypersurfaces of the systems HOH–F⁻ and HOH–Cl⁻.
Lower part: section in the plane of the water molecule.
Upper part: section perpendicular to the plane of the water molecule containing its C_{2v}-axis.
The spacing of the lines is 5.0 kcal/mole (analytical fits from Ref. [206])

F, the anion is bound to a water dimer, but occupying two different binding sites. Structures D and E are both stable against dissociation, whereas F is not. Because of the polarizing effect of F⁻, the hydrogen bond to the second water molecule is no longer efficient and complex F is found to dissociate spontaneously into H_2O and H_2O–F⁻. The corresponding structural data can be found in Table 23.

The two-dimensional density difference plot for structure D (Fig. 25) does not show any striking difference to the corresponding diagram for the one-to-one complex (Fig. 22). The degree of polarization of the ligands is reduced slightly due to the increased ion–ligand distance. The induced dipole moment in F⁻ is zero because of symmetry reasons. Using the pair-potential approximation, Monte-Carlo calculations have been made [162] on the dynamics of larger clusters simulating hydrated fluoride and chloride anions. At 0° K the preferred number of water molecules in the first solvation layer seems to be 4–6 for the fluoride and 6–7 for chloride anions. These rather broad distributions indicate that the structures differing in coordination number by one or two in the first shell have almost the same energy and the exchange of water molecules from the first to the second layer

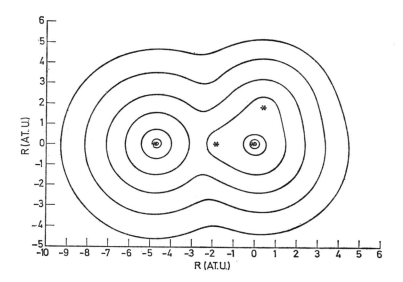

Fig. 22 a

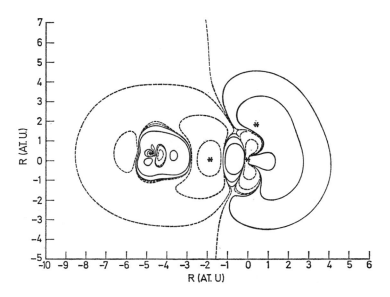

Fig. 22 b

Fig. 22. Total charge distribution and charge density differences upon formation of a HOH–F⁻ complex in the plane of the water molecule (maps are taken from Ref. [208])

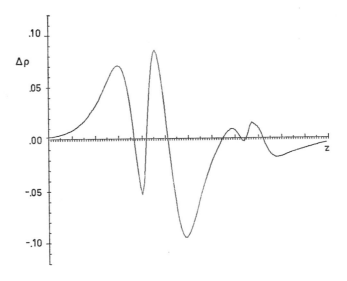

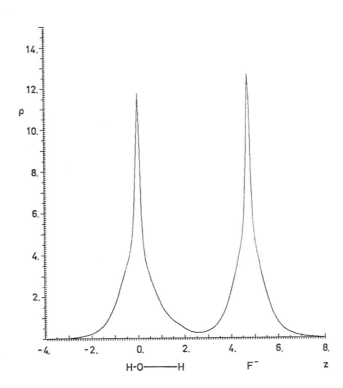

Fig. 23. Integrated charge density and integrated charge density differences in the system HOH–F⁻ complex. (Nuclear geometry from Ref. [203])

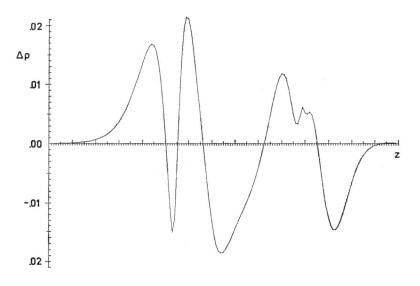

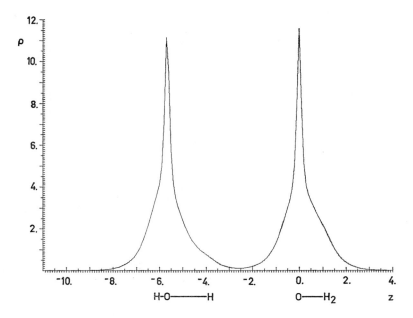

Fig. 24. Integrated charge density and integrated charge density differences in the water dimer (wave function taken from Ref. [211])

Table 23. *Ab initio* results for complexes of anions with several water molecules

Anion	Number of molecules	Ref.	Basis set[1]	Equilibrium geometry	Total stabilization energy, $-\Delta E$
F⁻	2	161)	GTO: F⁻O(11/7/1) [5/4/1] H(6/1) [3/1]	See Fig. 19D $R_{OF^-} = 2.55$	44.9
				See Fig. 19E $R_{OF^-} = 2.44$ $R_{OO} = 2.78$	36.6
				See Fig. 19F $R_{OF^-} = 2.52$ $R_{OO} = \infty$	24.7
F⁻	2	162)	GTO: F⁻(13/8/1) [7/3/1] O(11/7/1) [4/3/1] H(6/1) [2/1]	See Fig. 19D $R_{OF^-} = 2.55$	43.5[2])
F⁻	2	131)	—	—	39.9 exp. value
F⁻	3	162)	See two rows above	C₃ symmetric $R_{OF^-} = 2.58$	61.1[2])
F⁻	3	131)	—	—	53.6 exp. value
F⁻	4	162)	See four rows above	Tetrahedral $R_{OF^-} = 2.67$	73.4[2])
F⁻	4	131)	—	—	67.1 exp. value
Cl⁻	2	162)	GTO: Cl⁻(17/10/1) [9/6/1] O(11/7/1) [4/3/1] H(6/1) [2/1]	See Fig. 19D $R_{OCl^-} = 3.31$	22.9[2])
Cl⁻	2	131)	—	—	25.8 exp. value
Cl⁻	3	162)	see two rows above	C₃ symmetric $R_{OCl^-} = 3.34$	32.7[2])
Cl⁻	3	131)	—	—	37.5 exp. value

1) All lengths in (Å), energies in (kcal/mole).
2) See footnote 3) of Table 17, the ratio of the multibody effects are here 1./2.3/6.1/10.3 for F⁻-(H₂O)ₙ, n=3—6 and 1./2.4 for Cl⁻-(H₂O)ₙ, n=3—4.

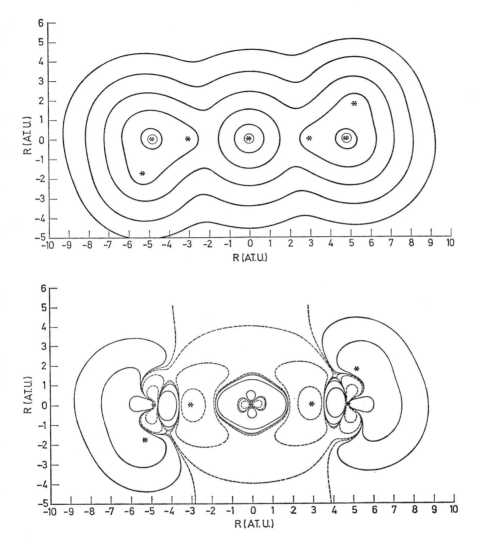

Fig. 25. Total charge density and charge-density differences in the HOH–F⁻–HOH complex in the plane of the water molecules (maps taken from Ref. [208])

can be performed in a cooperative way with very small activation energy. Correlation functions have also been calculated for the tetracoordinated anions F⁻ and Cl⁻ at room temperature ($T = 298\ °K$). Two maxima are observed in the ion hydrogen correlation function, roughly corresponding to both orientations discussed before (Fig. 19, $C\ \delta = 0°$, and $\delta = 52.3°$). As expected, the maxima of the distribution functions are much sharper in $F^-(H_2O)_4$ compared to $Cl^-(H_2O)_4$. Furthermore, the hydrogen-bonded structure is much more favored in the fluoride complex. These results are not likely to change appreciably for a larger number of ligands, as was demonstrated by the single calculation on $F^-(H_2O)_{27}$. Larger numbers of ligands lead to irregular structures that resemble a rudimentary second solvation layer. The destabilization which results if constraints are imposed on the complex geometry, like equal ion–oxygen distances, warns against direct comparison of solvation energies with stabilization energies calculated from rigid models. In order to test the validity of the pairwise additivity approximation Kistenmacher et al. [162] performed full H. F.-calculations on the systems $F^-(H_2O)_n$, $n = 1–5$, and $Cl^-(H_2O)_n$, $n = 1–3$. Many-body effects tend to destabilize the complexes. These effects are summarized in Table 23. Their relative magnitude are comparable to the values previously given for cation–water complexes.

2. Results Obtained with Semiempirical Procedures

CNDO/2 calculations performed on anion–water systems [222,223,225,234] show that the main deficiencies of this approach are much more pronounced than the results discussed previously for cations. At the CNDO-minimum geometries charge transfer reaches unrealistic high values (Table 24, Fig. 26). Binding energies are overemphasized by a factor of nearly four in the case of F^-HOH and by a factor of two in Cl^-HOH. Bond distances obtained systematically were too small (see Table 24). If, however, experimental or *ab initio* geometries are chosen for the

Table 24. Semiempirical (CNDO/II) calculations of anion–solvent interaction

Anion	Solvent	Ref.	Charge transfer to the ion	Cage geometry[1] R_{OX^-}	R_{OH_1}	Interaction[1] energy $-\Delta E$
F^-	1 (H_2O)	[222,223]	0.344	2.22	1.12	79.80
F^-	2 (H_2O)	[222,223]	0.410	2.29	1.09	118.30
F^-	2 (H_2O)	[234]	0.409	2.23	1.10	118.85
F^-	4 (H_2O)	[222,223]	0.452	2.36	1.06	164.00
F^-	4 (H_2O)	[234]	0.445	2.38	1.06	164.00
F^-	6 (H_2O)	[223]	0.440	2.43	1.05	189.30
F^-	6 (H_2O)	[234]	0.453	2.44	1.05	189.60
F^-	6 (H_2O)	[225]	—	2.34	—	192.00
F^-	8 (H_2O)	[223]	0.418	2.51	1.04	168.70
F^-	8 (H_2O)	[234]	0.423	2.53	1.04	195.20
F^-	8 (H_2O)	[225]	—	2.43	—	186.00
F^-	12 (H_2O)	[234]	0.430	2.72	1.04	186.40
Cl^-	1 (H_2O)	[222,223]	0.095	2.74	1.06	23.60
Cl^-	2 (H_2O)	[222,223]	0.146	2.76	1.05	45.60
Cl^-	4 (H_2O)	[222,223]	0.212	2.78	1.05	79.80
Cl^-	6 (H_2O)	[223]	0.225	2.81	1.04	108.30
Cl^-	6 (H_2O)	[225]	—	2.81	—	111.00
Cl^-	8 (H_2O)	[223]	0.264	2.84	1.04	98.20
Cl^-	8 (H_2O)	[225]	—	2.81	—	131.00
Cl^-	HCOOH	[230]	—	1.63/O–H..Cl^-		38.30
ClO_4^-	HCOOH	[235]	—	$R_{OO} = 2.70$ not varied		16.30

[1] All lengths in (Å), energies in (kcal/mole), all structures linear H-bonded.

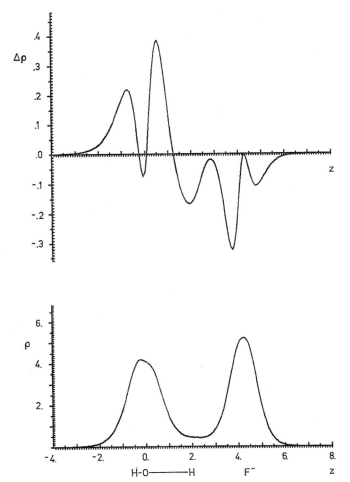

Fig. 26. Integrated charge density and integrated charge density differences of the HOH–F⁻ complex calculated in the CNDO/2 approximation (equilibrium geometry from Ref. [223])

semiempirical calculations the results improve substantially. Unfortunately these data are not usually known in advance.

Ligand–ligand interaction causes a decrease in incremental bond energies with increasing ligand number. In addition the total energy of interaction passes through a minimum at a coordination number of six for both anions F⁻ and Cl⁻. Inclusion of higher solvation shells does not change the preferred coordination number (six) for F⁻.

Besides water, other solvent molecules have been investigated to a small extent in connection with anion solvation. We mention here two examples of solvation by formic acid, concerning the ions Cl⁻ [230] and ClO$_4^-$ [235]. In the first case coordination numbers up to four were considered. For the one to one complexes the energy of interaction is larger for Cl⁻ than for ClO$_4^-$.

Table 25. E.H.M. calculations on anion–water interactions[1])

Ion	Number of H_2O molecules	Cage geometry[2])	Total stabilization energy[2]), $-\Delta E$
F[-]	4	Tetrahedral[3]) $R_{OF^-} = 2.34$	61.0
F[-]	6	Octahedral[3]) $R_{OF^-} = 2.38$	83.0
F[-]	8	Cubic[4]) $R_{OF^-} = 2.58$	167.0
Cl[-]	6	Octahedral[3]) $R_{OCl} = 2.63$	229.0
Cl[-]	8	Cubic[3]) $R_{OCl^-} = 2.72$	270.0

[1]) All data are from Ref. [225]).
[2]) Lengths in (Å), energies in (kcal/mole).
[3]) Linear H-bonded structures.
[4]) Ion–dipole oriented structure.

In general the semiempirical procedures permit qualitative insight at the molecular level if some precaution is taken, but quantitative results should be looked at with care and with the specific approximations of the method in mind.

VI. The Solvation of Polar Molecules

The structure of liquid mixtures and the interactions occurring in solutions of polar molecules are obviously even more difficult to describe than corresponding properties of pure liquids. Along with orientation- and aggregation processes, self association between solvent molecules or between solute molecules and the formation of mixed polymers are important constituents of the conceptual "basis-set", into which the global properties of a solution have to be expanded if one tries to envisage the highly complex events occurring in those solutions. Structural changes of the solute molecules due to internal geometrical relaxation caused by solute–solvent interactions must also be considered as well as phenomena like preferred interaction sites on the molecules. Such problems do not arise in the solvation of monatomic ions with their spherically symmetric appearance. There is also another reason that makes ionic solution easier to treat. The leading ion–dipole energy term in ion–molecule interactions overwhelms all the other contributions, which gives some significance to a static interpretation of ionic solutions in the discussion of ion–solvent clusters. No pronounced leading term exists in solutions of polar molecules, where all the different interactions might be of very similar magnitude. Certainly dynamical aspects are all-important in these solutions. We did not mention yet effects like solvolysis or acid–base reactions between solute and solvent, which might be quite prominent in solutions of polar molecules in polar liquids, and which would further obscure and complicate the whole solvation problem.

According to the complexity of the phenomenon of solvation, experimental results which might have bearings for its understanding are spread throughout the literature. Gas-phase investigations of intermolecular interactions are of a basic interest [236] and a large amount work has been done on the thermodynamics and spectroscopic properties of liquid mixtures [29,237–239].

Specific interactions between the polar solute and both polar or non-polar solvent molecules have been found to play a dominant role in solvation, particularly in the most outstanding specific interaction, hydrogen bonding. Remember for instance that hydrogen bonds between solute and solvent are traceable in solutions of H_2O in $CHCl_3$, as shown by near-IR investigations [240], and that theoretical calculations reveal H-bond-like interactions, *e.g.* between C_2H_4 and H_2O [241]. The vast field of hydrogen bonding is, however, so extensively documented and dicussed in the literature on intermolecular forces, that it is sufficient here to give a short reference to recent monographs dedicated to this problem [242,243].

Theoretical considerations based upon a molecular approach to solvation are not yet very sophisticated. As in the case of ionic solvation, but even more markedly, the connection between properties of liquid mixtures and models on the level of molecular colculations is, despite all the progress made, an essentially unsolved problem. Even very crude approximative approaches utilizing for example the concept of pairwise additivity of intermolecular forces are not yet tractable, simply because extended potential hypersurfaces of dimeric molecular associations are lacking. A complete hypersurface describing the potential of two diatomics has already a dimensionality of six! In this light, it is clear that advanced calculations are limited to very basic aspects of intermolecular interactions, and the use of less accurate methods is inevitable whenever one intends to go a little more towards "physical" solvation problems.

In accordance with the limited aim of this article, in the following subsections we shall consider only two of the individual contributions to the highly complex solvation process dealing with structural problems of solutions. No attempt is made to treat other aspects, which might well be of considerable theoretical interest, like electronic spectra [189,244] or dynamic properties [13] of liquid mixtures.

A. The Solute Molecules

Let us turn our attention first to the solute molecules themselves. The geometry of a molecule can certainly change upon solvation, compared to its appearance *in vacuo*. Thus an import first step towards a theoretical description of solutions of polar molecules concerns equilibrium geometries of solutes. There are some attempts which are not strictly "molecular" in our definition; however, they apply at least a complete Hamiltonian for the solute molecule. Incorporated into this Hamiltonian are the interactions with the solvent, which is then treated as a continuous dielectric, leaving a spherical cavity where the solute is placed. The equilibrium geometries of the dissolved molecule could be obtained as usual by looking for minima of the corresponding potential surface. As an additional merit, from the total energies with and without inclusion of the solvent terms, something like "solvation energies" are gained [185]. The method of Hylton *et al.* [185] has been mentioned already in Chapter IV.

Another attempt to find solute geometries without explicitly including solvent molecules in the calculations is due to Sinanoğlu [245,246]. In a recent paper he proposed a "C-potential" effective for molecules in solution, which is derived from the potential surface of a naked solute molecule by inclusion of additive solvation terms obtainable from simple macroscopic properties of the pure liquid solvent. This method is an extension of an earlier formalism applicable to intermolecular potentials between solvated molecules [247,248].

In a more systematic way, geometries of interest could in principle be derived from minima of the quantum mechanical potential energy hypersurfaces of clusters including solvent molecules. To yield full quantitative information for physical solutions, a cluster big enough to satisfactorily represent the conditions of solutions would have to be considered. The computational treatment of such a complex is obviously an impossible exercise. A restriction to smaller model-clusters, the use of frozen geometries for solvent molecules and of fixed intermolecular arrangements as well as a proper selection of some geometrical parameters of the solute would bring the problem within the manageable size of semiempirical quantum chemical methods. Using this supermolecule approach and the PCILO-method [249–251], Pullman and co-workers [252,253] have recently investigated structural changes in solution for some molecules of pharmacological importance, like the histamine cation. Potential energy curves for variations of some bond angles of interest were eventually found to be quite different for unhydrated or hydrated molecules.

Clearly, with increasing strength of interaction between solute and solvent the strain acting on the molecules to change their geometries in creases. Furthermore, with increasing size of the molecules as well as with increasing number of interaction sites the flexibility and hence the possibility for geometrical relaxation of the molecule increases. For instance, closed dimeric structures show a greater tendency of geometrical relaxation than open ones [18]. Accordingly, we might expect more drastic geometrical variations in solutions of large molecules than in those of smaller ones.

B. Solute-Solvent Interactions

In theoretical investigations of molecular solute–solvent interactions, the method of molecular electrostatic potentials [19] (cf. Chapters II and IV) has proved to be a very useful tool for a quick search of preferred solvations sites [50–54]. More rigorous quantum mechanical computations concerning molecular complexes have almost exclusively been dedicated to hydrogen-bonded systems. This confinement is not surprising. With reference to very time-consuming accurate *ab initio* calculations, a good "first guess" as to the geometrical arrangment in the complex to be calculated would certainly be well received. A large amount of semiquantitative experience permits a reasonable first guess in the case of hydrogen-bonded systems. Furthermore, semiempirical methods like for example CNDO/2 are well known to be rather more reliable to hydrogen bonds if used carefully and critically, than to other types of intermolecular interactions [18]. These facts have also channelled research work preferably into the hydrogen-bond business.

Theoretical treatments and calculations of hydrogen-bonded systems have been reviewed thoroughly recently [16–18]. We would like to indicate only a few examples of *ab initio* calculations here. *Ab initio* investigations are restricted mainly to dimeric structures, whereby on this level it might be a matter of definition which of the partners is regarded to be the solute molecule and which the solvent. Besides, most investigations deal with very limited areas of the potential hypersurface, often employing fixed geometries within the subunits and presenting results only in the vicinity of the stable hydrogen-bonded conformation. For a list of the systems treated so far see [18].

There are only a few calculations on hetero-dimers employing large basis sets. Diercksen *et al*. [254] in their calculations on the NH_3–H_2O system have used an $(11, 7, 1; 6, 1)$ basis set not yet surpassed. Both the structures with either H_2O or NH_3 as proton acceptor have been considered. Obviously, H_3N–HOH is preferred over H_2O-HNH_2, NH_3 being the better H-acceptor and in turn H_2O being the better H-donor (Table 26). Dimeric water and dimeric ammonia are weaker complexes than the mixed dimer. The differences, however, are not very pronounced, and no predictions can be made concerning a dominance of hetero-structures in aqueous solutions, where the ability of the water molecules to form extended hydrogen-bonded structures must be taken into account. In this connection, nonadditivity effects in hydrogen bonding [256] should be pointed out once more. Only in dilute gases we may expect conclusively to find mixed polymers.

Table 26. Dimeric complexes in H_2O-NH_3 mixtures. Results of *ab initio* calculations

System	Basis set	R_{XY} (Å)	ΔE (kcal/mole)
H_3N-HOH[1]	(11,7,1; 6,1)	3,06	−6.28
H_2O-HNH_2[1]	(11,7,1; 6,1)	3.35	−2.33
H_2O-HOH[2]	(11,7,1; 6,1)	3,00	−4.83
H_3N-HNH_2[3]	(10,5; 5)	3.49	−2.71

[1] Ref. [254].   [2] Ref. [100].   [3] Ref. [255].

Another system to be explicitly discussed here concerns the hydration of formaldehyde. Morokuma's calculations [85] with a minimal Slater basis set provide a rare example of an *ab initio* investigation on a trimeric structure, which might already serve as a crude model for the hydration of the carbonyl group. The energy surface for the dimer H_2CO-H_2O was covered by Morokuma [85] also more extensively than usual. As expected, the most stable dimeric structure is characterized by a linear hydrogen bond directed toward a lone pair on the carbonyl oxygen. The binding energy is calculated as 3.45 kcal/mole. (Using the same basis set, the hydrogen bond energy of the water dimer is found to be 6.55 kcal/mole [257].) The dimeric complex does not lose much of its stability when the hydrogen bonds are distorted away from linearity by as much as $\pm15°$. In addition, the potential hypersurface also reveals bound regions for hydrogen bonds pointing towards the π-system of formaldehyde and for structures with H_2O as the proton acceptor.

Recently, Morokuma *et al.* [244] extended their original calculation to include excited states, using larger GTO basis sets. For reasons of comparison they repeated their calculations on the ground state of the complex, replacing the minimal Slater basis by the STO-3G representation [92] and also applied the larger basis set designated 4-31G by Ditchfield *et al.* [258]. In addition, both basis sets were also used augmented with a set of diffuse p-orbitals proposed by Whitten [259] for the calculation of excited states. As in the case of $Li^+.OH_2$, the resulting binding energies are very sensitive to an increase in the number of basis functions. The interaction energy obtained with the larger 4-31G basis set is remarkably higher than the one computed with the minimal basis: 6.3 kcal/mole in place of 3.4 kcal/mole. Energy partitioning, as described extensively in Chapter II. [85], reveals that this increase is mainly due to electrostatic contribution (reflecting the fact that the dipole moments of both H_2CO and H_2O are underestimated by the minimal basis, but overestimated by the larger basis set). A pronounced decrease of the charge transfer term similar to the case of $Li^+.OH_2$ (Table 3) cannot fully compensate the electrostatic effect. A comparatively small contribution to the enlargement of the hydrogen bond energy is provided by the plolarization term (Table 27).

The geometry of the most stable dimeric structure suggests that a second water molecule might be bound to the carbonyl group symmetrically to the first one, utilizing the second "lone pair". Using geometrical input data similar to those for the optimized dimer, a stabilization energy of 2.50 kcal/mole is gained for the second hydrogen bond, roughly 70% of the energy for the first bond. (This value might change a little on optimization of the trimer's geometry.) An analysis of the

Table 27. Intermolecular energy partitioning in $H_2CO \cdot HOH$ obtained with different basis sets

Quantity[1])	STO [85]) min. basis	Basis set STO–3 G [244])	4–31 G [244])
μ_{H2O} [D]	—	1.71	2.67
μ_{H2CO} [D]	—	1.53	3.01
ΔE_{COU}	−4.6	−4.60	−9.70
ΔE_{EX}	6.7	7.30	6.90
ΔE_{POL}	−0.2	−0.10	−0.80
ΔE_{CHT}	−5.3	−5.90	−2.80
ΔE	−3.4	−3.40	−6.30

[1]) Energy differences in (kcal/mole).

nonadditivity effects, according to Hankins et al. [89], shows that the three-body term is particularly small in this complex, being a repulsive contribution of only 0.05 kcal/mole.

More water molecules would have to go to less favorable positions above the π-bond of H_2CO, or preferably, would attach to the water molecules in the complex. The trimeric structure could therefore hopefully serve as a model for the first hydration shell of the carbonyl group, at least in the gas phase. Again in the liquid phase we must not expect a prevalence of these structures, since the small stabilization energies (only few kilocalories/mole) can easily be overcompensated by water–water interactions. In addition it should be recalled that aldehydes and ketones are able in principle to undergo a chemical reaction with water in aqueous solutions to form compounds $R_1R_2C(OH)_2$, which are known as "hydrates" in organic chemistry [260]. The latter are distinct chemical entities whose hydration properties would have to be studied separately. Formaldehyde is particularly accessible to this process. In aqueous solution at room temperature it exists to 99.99% as "hydrate". Both electron-donating or bulky substituents will diminish the tendency to chemical hydration of the keto-group. Aqueous acetaldehyde at 293°K has an equilibrium composition of 58% "hydrate", and acetone does not form an appreciable amount of "hydrate" under these conditions [261]. Thus formaldehyde is not a good object for a discussion of solvation problems in water, but rather a convenient model for the quantum chemist.

Supplementary to the data for H_2CO–H_2O complexes are results derived by Del Bene [262], who also used a minimal basis set to calculate stabilization energies and geometries for the series H_2CO–HOR, where R is one of a number of increasingly electron-withdrawing substituents, namely H, CH_3, NH_2, OH and F. For H_2O, the binding energy of 3.33 kcal/mole is in good agreement with Morokuma's value (3.45 kcal/mole [85]). Hydrogen bonds become stronger in this series, reaching a value of 5.00 kcal/mole for FOH (Table 28). As expected, the increase in hydrogen bond strength reflects the increasing acidity of the respective proton. Additionally, the "solvent" molecules CH_3 to FOH, will not be able to form extended networks of hydrogen bonds like water, and thus trimeric complexes similar to the one proposed by Morokuma [85] become more likely candidates to be actually observed in corresponding solutions.

There is a big step from the structure of solutions to the treatment of reactions in solution. To close this chapter, investigations dealing with the evaluation of

Table 28. H-bonding in complexes $H_2CO\text{–}HOR$. Calculations with minimal basis sets

R	R_{OO} (Å)		ΔE(kcal/mole)	
H	2.87[1])	2.88[2])	−3.45[1])	−3.33[2])
CH_3		2.86[2])	−3.49[2])	
NH_2		2.80[2])	−4.19[2])	
OH		2.78[2])	−4.39[2])	
F		2.77[2])	−5.00[2])	

[1]) Ref. [85].
[2]) Ref. [262].

reaction pathways including solvent molecules shall be noted. In fact, these calculations also deal with hydrogen-bonded solute–solvent systems, which might permit a meaningful model of the complexes involved to be made. Cremaschi *et al.* [234] performed CNDO/2 calculations on the S_N2 inversion reaction

$$F^- + CH_3F \longrightarrow [CH_3F_2]^- \longrightarrow FCH_3 + F^- \tag{51}$$

where all the individual entities involved were treated as hydrated complexes, including just a "first hydration shell". Schuster, *et al.* [263] investigated the influence of a hydration shell on complex formation and proton transfer in $NH_3 . HF$

$$NH_3 + HF \longrightarrow NH_3 . HF \longrightarrow [NH_4]^+ . F^- \tag{52}$$

also using the CNDO/2 method. According to these calculations, hydration tends to stabilize ionic species more than neutral molecules or complexes, thus facilitating the proton transfer.

It should be pointed out that one cannot expect quantitatively correct data from such calculations. Clearly, the complexes considered do not appropriately represent real solutions. Most of the results obtained could have been guessed equally well by chemical experience and intuition: anyway we expect ions to be more strongly hydrated than neutral molecules. In the actual calculations, the method employed is known to overemphasize the expected effects. The merits of attempts like the ones mentioned are therefore not to be found in the realization of quantitative results, but verify that our expectations are definitely reproducable in terms of quantum chemical data, and they demonstrate how such calculations could be made. There have also been attempts to describe reactions of solvated molecules by an MO theoretical treatment for the two reaction partners, with inclusion of the solvent by representing it as point dipoles. As a first step, Yamabe *et al.* [186] performed *ab initio* calculations on the complex $NH_3 . HF$, "solvating" each of the partners by just one point dipole. A study of MO's of the interacting complex with and without dipoles shows that the latter has a favorable effect on the proceeding of the reaction.

Summarizing this chapter briefly, available calculations generally concern very limited aspects of the solvation of polar molecules. We have to be aware of the fact that much remains to be learnt about the theoretical treatment of associated liquids and liquid mixtures.

VII. Conclusion and Further Developments

The results of various attempts to calculate interactions between ions and solvent molecules which have been summarized in this article suggest that a degree of accuracy can be achieved nowadays, which is sufficient for most purposes. Reliable intermolecular geometries and energies of interaction can be obtained from *ab initio* calculations at the H. F. level and in most cases with medium-sized basis sets. Changes in properties of ligands brought about by the interaction with the solute particle and their consequences in molecular spectra are much more difficult to predict. Great care is needed in order to make sure that the calculated results are free from methodical artifacts. Furthermore, electron correlation effects might be important in this case as well. Nevertheless, it seems that accurate two-body potentials are accesible now with reasonable efforts, providing the interacting particles are small enough. Consequently, the interest of most theoreticians working in this field has been shifted already towards three-body potentials and higher contributions to the energy surfaces of clusters.

Until now, ions of the alkali- and alkaline earth series and halogenid ions have been studied almost exclusively. A consequent extension of the work performed in this field would lead inevitably to systematic solvation studies of transition-metal ions or to investigations of various types of polyatomic anions. For a long period, complexes of transition-metal ions have been the subject of interest with different goals in mind. Approximate methods and model theories, mainly crystal field theory [264,265] and simple empirical MO methods were applied to the study of for example electron absorption spectra or magnetic properties of these complexions. Some semiempirical MO methods of the CNDO type were also used in calculations of transition-metal complexes at frozen intermolecular geometries [266]. In general, however, all these methods were found to be either inadequate or not very useful in calculations of energy surfaces for solvent–ion interactions. On the other hand, in these complexes the number of electrons is usually so large that accurate MO calculations are beyond our present limits of computational techniques. Recently developed multiple scattering methods, *e.g.* [267,268], might provide an appropriate tool for investigations of these structures.

Calculations on dynamics of solvation shells are still in their infancy. However, very recent papers on this subject, show that in most examples we cannot expect a realistic picture of solvent shells from a purely static approach. Most probably, molecular dynamics calculations and Monte Carlo methods will produce a variety of interesting data and will improve our knowledge on solvation of ions substantially.

As a promising starting point for theoretical work there is now an increasing wealth of very detailed experimental data concerning solvation. Apart from crude overall properties as known from the thermodynamical measurements we are now in a position to gain energetic descriptions for the solvation process itself by gasphase solvation experiments without any perturbation from the liquid phase. Sophisticated measurements for liquids might reveal detailed average values for static and dynamic aspects of solutions, *e.g.* average geometries and residence times of associations. A combination of the theoretical and experimental results now available puts a narrow restriction on the construction of meaningful models, and the theories of solvation tend to converge towards a realistic picture of ions in solution.

In the case of polar molecules the gap between theory and reality is even greater, mainly due to two reasons. Since the interacting particles are usually greater in size but less symmetric, theoretical calculations are much more time-consuming and only less accurate methods can be applied. Secondly, the forces acting between the particles are weaker and therefore long for more elaborate techniques in both static and dynamic calculations. Similar problems occur in phenomenological theories and even in the experimental approach. For a full description of solvation shells of polar molecules more information would be necessary than in the case of ions but less experimental data are available. On the whole, our knowledge of the structure of clusters formed from polar molecules in solution or in liquid mixtures is by far inferior to that of solvation of ions.

Note Added in Proof: After we sent the manuscript to the publishers we became aware of CNDO studies on alkali ion solvation performed by Gupta and Rao [270] and Balasubramanian *et al.*[271], which might be of some importance for readers interested in cation solvation by water and various amides. Another CNDO model investigation on the structure of hydrated ions was published very recently by Cremaschi and Simonetta [272]: They studied CH_5^+ and CH_5^- surrounded by a first shell of water molecules in order to discuss solvation effects on structure and stability of these organic intermediates or transition states respectively.

Accurate *ab initio* studies on one to one complexes of oneatomic ions and water were performed as well. In the final version of their very extensive study Diercksen *et al.*[108] presented large scale CI calculations. As expected, the contribution of electron correlation to the energies of complex formation is rather small, destabilizing in $Li^+ \ldots OH_2$ ($\Delta E_{COR} = 1.2$ kcal/mole), stabilizing in $F^- \ldots$ HOH ($\Delta E_{COR} = -9.1$ kcal/mole) and $(H_2O)_2$ ($\Delta E_{COR} = -0.9$ kcal/mole). Recent work of our group [91,109] was dealing with the basis set dependence of SCF energy partitioning and simple electrostatic expressions were found to account reasonably well for the contributions of ΔE_{COU} and ΔE_{POL} to total energies of interaction. Another interesting paper was presented by Chan *et al.*[273], who computed the whole dipole moment surface of the $Li^+.OH_2.F^-$ complex.

During the last year two Monte Carlo calculations and one molecular dynamics study [276] on ion hydration were published. Watts *et al.*[274] extended previous Monte Carlo calculations [162] to the Li^+F^- ion pair, which was surrounded in this model calculation by 50 water molecules. In the more recent work of this group Fromm *et al.*[275] increased the number of solvent molecules to 200 and coordination numbers for the ions at room temperature were evaluated ($n_{Li^+} \sim 6$, $n_{F^-} \sim 4$). Vogel and Heinzinger [276] extended their molecular dynamics study on LiCl in H_2O [183] to an aqueous CsCl solution.

Acknowledgements:

The authors express their gratitude to Drs. G. H. F. Diercksen and W. P. Kraemer for sending density and density difference plots of $Li^+(OH_2)_n$ and $F^-(HOH)_n$ prior to publication and for the permission to print them in this review article. Several other groups sent their most recent results prior to publication. We want to thank all of them. Numerical calculations of our group reported here were performed on an IBM 360/44 (Interfakultäres Rechenzentrum der Univ., Vienna), a UNIVAC 1108 (Gesellschaft für Wissenschaftliche Datenverarbeitung Göttingen), an IBM 1130 (Hochschule für Bodenkultur, Vienna) and an IBM 370/165/C.I.R. C.E. (Paris Sud, Orsay). We are indebted to all these computing centers for their generous supply of computer time. The assistance of our technical staff, Mrs. J. Dura and Mr. J. Schuster with the preparation of the manuscript is gratefully acknowledged.

VIII. References

1) Frank, H. S., Wen, W.-Y.: Discussions Faraday Soc. *24*, 133 (1957).
2) Gurney, R. W.: Ionic processes in solution. New York: Dover Publ. Co. 1962.
3 Hertz, H. G., Lindman, B., Siepe, V.: Ber. Bunsenges. Physik. Chem. *73*, 542 (1969).
4) Bernal, J. D., Fowler, R. H.: J. Chem. Phys. *1*, 515 (1933).
5) Verwey, E. J. W.: Rec. Trav. Chim. *61*, 127 (1942).
6) Morf, W. E., Simon, W.: Helv, Chim. Acta *54*, 794 (1971).
7) McMillan, W. G., Mayer, J. E.: J. Chem. Phys. *13*, 276 (1945).
8) Friedman, H. L.: Mol. Phys. *2*, 23 (1959).
9) Jones, R. W., Mohling, F.: J. Phys. Chem. *75*, 3790 (1971).
10) Stecki, J.: Ionic solvation. In Ref. [236], pp. 413—458.
11) Stillinger, F. H., Rahman, A. J.: Chem. Phys. *60*, 1545 (1974).
12) Larsen, B., Førland, T., Singer, K.: Mol. Phys. *26*, 1521 (1973).
13) McDonald, I. R., Singer, K.: Chem. Brit. *9*, 54 (1973).
14) Murthy, A. S. N., Rao, C. N. R.: J. Mol. Struct. *6*, 253 (1970).
15) Pimentel, G. C., McClellan: Ann. Rev. Phys. Chem. *22*, 347 (1971).
16) Kollman, P. A., Allen, L. C.: Chem. Rev. *72*, 283 (1972).
17) Schuster, P.: Z. Chem. *13*, 41 (1973).
18) Schuster, P.: Energy surfaces for hydrogen bonded systems. In Ref. [242].
19) Scrocco, E., Tomasi, J.: The electrostatic molecular potential as a tool for the interpretation of molecular properties. Topics Curr. Chem. *42*, 95—170 (1973).
20) Morris, D. F. C.: Struct. Bonding *4*, 63 (1968) and ibid. *6*, 157 (1969).
21) Mayer, U., Gutmann, V.: Struct. Bonding *12*, 113 (1972).
22) Rosseinsky, D. R.: Chem. Rev. *65*, 467 (1965).
23) Hertz, H. G.: Angew. Chem. Intern. Ed. Engl. *82*, 91 (1970).
24) Friedman, H. L.: Chem. Brit. *9*, 300 (1973).
25) Bockris, J. O'M., Reddy, A. K. N.: Modern electrochemistry, Vol. *1*, pp. 45—174. New York: Plenum Press 1970.
26) Coetzee, J. F., Ritchie, C. D. (ed.): Solute–solvent interactions. New York: Dekker 1969.
27) Friedman, H. L., Krishnan, C. V.: Thermodynamics of ion hydration. In Ref. [29], Vol. *3*, pp. 1—118.
28) Conway, B. E., Barradas, R. G. (eds.): Chemical physics of ionic solutions. New York: J. Wiley 1966.
29) Franks, F. (ed.): Water, Vol. *1—4*. New York: Plenum Press 1973.
30) Horne, R. A. (ed.): Water and aqueous solutions. New York: Wiley-Interscience 1972.
31) Amis, E. S.: Solvent effects on reaction rates and mechanisms. New York: Academic Press 1966.
32) Hinton, J. F., Amis, E. S.: Chem. Rev. *67*, 367 (1967).
33) Hinton, J. F., Amis, E. S.: Chem. Rev. *71*, 627 (1971).
34) Burgess, J., Symons, M. C. R.: Quart. Rev. *22*, 276 (1968).
35) Blandamer, M. J., Fox, M. F.: Chem. Rev. *70*, 59 (1970).
36) Blandamer, M. J.: Quart Rev. *24*, 169 (1970).
37) Grunwald, E., Ralph, E. K.: Accounts Chem. Res. *4*, 107 (1971).
38) Case, B.: Ion solvation. In: Reactions of molecules at electrodes, pp. 46—134 (ed. N. S. Hush). New York: J. Wiley 1971.
39) Kecki, Z.: Advan. Mol. Relax. Proc. *5*, 137 (1973).
40) Conway, B. E.: Ann. Rev. Phys. Chem. *17*, 481 (1966).
41) Rosseinsky, D. R.: Ann. Rep. Chem. Soc. (London) *A 68*, 81 (1971).
42) Hills, G. J. (ed.): Electrochemistry (Specialist periodical reports). Chem. Soc. (London), Vol. *1* (1970) and Vol. *2* (1972).
43) Bockris, J. O., Conway, B. E. (eds.): Modern aspects of electrochemistry, Vol. *1*—Vol. *6*.
44) Bockris, J. O. (ed.): Electrochemistry, MTP International Review of Science Ser. 1, Vol. *6*. London: Butterworth 1972.
45) Hirschfelder, J. O., Curtiss, C. F., Bird, R. B.: Molecular theory of gases and liquids, pp. 835—1103. New York: J. Wiley 1954.

46) Margenau, H., Kestner, N. R.: Theory of intermolecular forces, 2nd. ed., pp. 15—298. Oxford: Pergamon Press 1971.
47) Hirschfelder, J. O., Meath, W. J.: The nature of intermolecular Forces (ed. J. O. Hirschfelder). Advan. Chem. Phys. *12*, 3—106 (1967).
48) Hagler, A. T., Lifson, S.: J. Am. Chem. Soc., in press.
49) Hagler, A. T., Huler, E., Lifson, S.: J. Am. Chem. Soc., in press.
50) Alagona, G., Pullman, A., Scrocco, E., Tomasi, J.: Intern. J. Peptide Protein Res. *5*, 251 (1973).
51) Pullman, A., Port, G. N. J.: Compt. Rend. Acad. Sci. Paris D *227*, 2269 (1973).
52) Port, G. N. J., Pullman, A.: FEBS Letters *31*, 70 (1973).
53) Port, G. N. J., Pullman, A.: Theoret. Chim. Acta *31*, 231 (1973).
54) Pullman, A., Alagona, G., Tomasi, J.: Theoret. Chim. Acta *33*, 87 (1974).
55) Kauzmann, W.: Quantum chemistry, p. 729. New York: Academic Press 1957.
56) Kollman, P.: J. Am. Chem. Soc. *94*, 1837 (1972).
57) Schaefer, H. F., III.: The electronic structure of atoms and molecules. Reading, Mass.: Addison-Wesley 1972.
58) Kutzelnigg, W.: Electron correlation and electron pair theories. Topics Curr. Chem. *41*, 31—73 (1973).
59) Pople, J. A., Santry, D. P., Segal, G. A.: J. Chem. Phys. *43*, S 129 (1965).
60) Pople, J. A., Segal, G. A.: J. Chem. Phys. *43*, S 136 (1965).
61) Pople, J. A., Segal, G. A.: J. Chem. Phys. *44*, 3289 (1966).
62) Pople, J. A., Beveridge, D. L.: Approximate molecular orbital theory. New York: McGraw Hill 1970.
63) Murrell, J. N., Harget, A. J.: Semiempirical self-consistent-field molecular orbital theory of molecules. London: Wiley-Interscience 1972.
64) Wolfsberg, M., Helmholtz, L.: J. Chem. Phys. *20*, 837 (1952).
65) Hoffmann, R.: J. Chem. Phys. *39*, 1397 (1963).
66) Schuster, P.: Monatsh. Chem. *100*, 1033 (1969).
67) Eisenschitz, H., London, F.: Z. Physik *60*, 491 (1930).
68) Chipman, D. M., Bowman, J. D., Hirschfelder, J. O.: J. Chem. Phys. *59*, 2830 (1973).
69) Chipman, D. M., Hirschfelder, J. O.: J. Chem. Phys. *59*, 2838 (1973).
70) Daudey, J. P., Claverie, P., Malrieu, J. P.: Intern. J. Quantum Chem. *8*, 1 (1974).
71) Daudey, J. P., Malrieu, J. P., Rojas, O.: Intern. J. Quantum Chem. *8*, 17 (1974).
72) Daudey, J. P.: Intern. J. Quantum Chem. *8*, 29 (1974).
73) Murrell, J. N., Randić, M., Williams, D. R.: Proc. Roy. Soc. (London) A *284*, 566 (1965).
74) Murrell, J. N., Shaw, G.: J. Chem. Phys. *46*, 1768 (1967).
75) Murrell, J. N., Teixeira-Dias, J. J. C.: Mol. Phys. *19*, 521 (1970).
76) Conway, A., Murrell, J. N.: Mol. Phys. *23*, 1143 (1972).
77) Murrell, J. N., Shaw, G.: Mol. Phys. *12*, 475 (1967).
78) Van Duijneveldt-Van de Rijdt, J. G. C. M., Van Duijneveldt, F. B.: Chem. Phys. Letters *17*, 425 (1972).
79) Williams, D. R., Schaad, L. J., Murrell, J. N.: J. Chem. Phys. *47*, 4916 (1967).
80) Van Duijneveldt-Van de Rijdt, J. G. C. M., Van Duijneveldt, F. B.: J. Am. Chem. Soc. *93*, 5644 (1971).
81) Schuster, P.: Chem. Phys. Letters, in preparation.
82) Lischka, H.: J. Am. Chem. Soc. *96*, 4761 (1974).
83) Dreyfus, M., Pullman, A.: Theoret. Chim. Acta *19*, 20 (1970).
84) Kollman, P. A., Allen, L. C.: Theoret. Chim. Acta *18*, 399 (1970).
85) Morokuma, K.: J. Chem. Phys. *55*, 1236 (1971).
86) Clementi, E.: J. Chem. Phys. *46*, 3842 (1967).
87) Clementi, E., Popkie, H.: J. Chem. Phys. *57*, 1077 (1972).
88) Popkie, H., Kistenmacher, H., Clementi, E.: J. Chem. Phys. *59*, 1325 (1973).
89) Hankins, D., Moskowitz, J. W., Stillinger, F. H.: J. Chem. Phys. *53*, 4544 (1970) and erratum *ibid. 59*, 995 (1973).
90) Kollman, P. A., Kuntz, I. D.: J. Am. Chem. Soc. *94*, 9236 (1972).
91) Beyer, A., Lischka, H., Schuster, P.: to be published.

92) Hehre, W. J., Stewart, R. F., Pople, J. A.: J. Chem. Phys. 51, 2657 (1969).
93) Coulson, C. A., Danielson, U.: Arkiv Fysik 8, 205 (1955).
94) Coulson, C. A.: Research 10, 149 (1957).
95) Buckingham, A. D.: Discussions Faraday Soc. 24, 151 (1957).
96) Buckingham, A. D.: Quart. Rev. (London) 13, 189 (1959).
97) Buckingham, A. D.: Permanent and induced molecular moments and long-range intermolecular forces. In Ref. 236), pp. 107—142.
98) Nelson, R. D., Jr., Lide, D. R., Jr., Maryott, A. A. (eds.): Selected values of dipole
moments for molecules in the gas phase. Washington, D. C.: Nat. Bureau of Standards
1967.
99) Verhoeven, J., Dymanus, A.: J. Chem. Phys. 52, 3222 (1970).
100) Neumann, D., Moskowitz, J. W.: J. Chem. Phys. 49, 2056 (1968).
101) Hüttner, W., Lo, M. K., Flygare, W. H.: J. Chem. Phys. 48, 1206 (1968).
102) Neumann, D., Moskowitz, J. W.: J. Chem. Phys. 50, 2216 (1969).
103) Landolt, H. H., Börnstein, R.: Zahlenwerte und Funktionen, Vol. 1, pp. 510. Berlin–
Göttingen–Heidelberg: Springer 1951.
104) Liebmann, S. P., Moskowitz, J. W.: J. Chem. Phys. 54, 3622 (1971).
105) Spears, K. G.: J. Chem. Phys. 57, 1850 (1972).
106) Rizzi, A., Lischka, H., Schuster, P.: to be published.
107) Buckingham, A. D.: Discussions Faraday Soc. 40, 232 (1965).
108) Diercksen, G. H. F., Kraemer, W. P., Roos, B. O.: Theoret. Chim. Acta 36, 249 (1975).
109) Marius, W., Schuster, P., Pullman, A., Berthod, H.: in preparation.
110) Pullman, A., Schuster, P.: Chem. Phys. Letters 24, 472 (1974).
111) Schuster, P., Preuss, H. W.: Chem. Phys. Letters 11, 35 (1971).
112) Mulliken, R. S.: J. Chem. Phys. 23, 1833, 1841, 2338 and 2343 (1955).
113) Russegger, P., Schuster, P.: Chem. Phys. Letters 19, 254 (1973).
114) Born, M.: Z. Physik 1, 45 (1920).
115) Stern, K. H., Amis, E. S.: Chem. Rev. 59, 1 (1959).
116) McDaniel, E. W., Čermák, V., Dalgarno, A., Ferguson, E. E., Friedman, L.: Ion–molecule
reactions. New York: Wiley-Interscience 1970.
117) Franklin, J. L. (ed.): Ion–molecule reactions. New York: Plenum Press, 1972.
118) De Paz, M., Leventhal, J. J., Friedman, L.: J. Chem. Phys. 51, 3748 (1969).
119) Kebarle, P., Godbole, E. W.: J. Chem. Phys. 39, 1131 (1963).
120) Kebarle, P., Hogg, A. M.: J. Chem. Phys. 42, 798 (1965).
121) Hogg, A. M., Kebarle, P.: J. Chem. Phys. 43, 449 (1965).
122) Hogg, A. M., Haynes, R. M., Kebarle, P.: J. Am. Chem. Soc. 88, 28 (1966).
123) Kebarle, P., Haynes, R. M., Collins, G. J.: J. Am. Chem. Soc. 89, 5753 (1967).
124) Kebarle, P., Searles, S. K., Zolla, A., Scarborough, J., Arshadi, M.: J. Am. Chem. Soc. 89,
6393 (1967).
125) Kebarle, P., Arshadi, M., Scarborough, J.: J. Chem. Phys. 49, 817 (1968).
126) Kebarle, P.: Advan. Chem. Ser. 72, 24 (1968).
127) Searles, S. K., Kebarle, P.: J. Phys. Chem. 72, 742 (1968).
128) Searles, S. K., Kebarle, P.: Can. J. Chem. 47, 2619 (1969).
129) Good, A., Durden, D. A., Kebarle, P.: J. Chem. Phys. 52, 212 (1970).
130) Džidič, I., Kebarle, P.: J. Phys. Chem. 74, 1466 (1970).
131) Arshadi, M., Yamdagni, R., Kebarle, P.: J. Phys. Chem. 74, 1475 (1970).
132) Arshadi, M., Kebarle, P.: J. Phys. Chem. 74, 1483 (1970).
133) Kebarle, P.: J. Chem. Phys. 53, 2129 (1970).
134) Yamdagni, R., Kebarle, P.: J. Am. Chem. Soc. 93, 7139 (1971).
135) Yamdagni, R., Kebarle, P.: J. Am. Chem. Soc. 94, 2940 (1972).
136) Kebarle, P.: Ions and ion–solvent molecule interaction in the gas phase. In: Szwarc, M.
(ed.), Ions and ion pairs in organic reactions, Vol. 1, pp. 27—83. New York: Wiley-
Interscience 1972.
137) Cunningham, A. J., Payzant, J. D., Kebarle, P.: J. Am. Chem. Soc. 94, 7627 (1972).
138) Yamdagni, R., Kebarle, P.: J. Am. Chem. Soc. 95, 3504 (1973).
139) Yamdagni, R., Payzant, J. D., Kebarle, P.: Can. J. Chem. 51, 2507 (1973).
140) Payzant, J. D., Cunningham, A. J., Kebarle, P.: Can. J. Chem. 51, 3242 (1973).

141) Grimsrud, E. P., Kebarle, P.: J. Am. Chem. Soc. 95, 7939 (1973).
142) Hiraoka, K., Grimsrud, E. P., Kebarle, P.: J. Am. Chem. Soc. 96, 3359 (1974).
143) Yamdagni, R., Kebarle, P.: Can. J. Chem. 52, 2449 (1974).
144) Durden, D. A., Kebarle, P., Good, A.: J. Chem. Phys. 50, 805 (1969).
145) Kebarle, P.: Higher-order reactions-ions cluster and ion solvation. In: Ref. 116), Vol. 1, pp. 315—362.
146) Futrell, J. H., Tiernan, T. O.: Tandem mass spectrometric studies of ion—molecule reactions. In: Ref. 117), Vol. 2, pp. 485—551.
147) Ferguson, E. E.: Flowing afterglow studies. In: Ref. 117), Vol. 2, pp. 363—393.
148) Fehsenfeld, F. C., Ferguson, E. E.: J. Chem. Phys. 59, 6272 (1973).
149) De Paz, M., Leventhal, J. J., Friedman, L.: J. Chem. Phys. 49, 5543 (1968).
150) Fehsenfeld, C., Mosesman, M., Ferguson, E. E.: J. Chem. Phys. 55, 2115 (1970).
151) Fehsenfeld, F. C., Mosesman, M., Ferguson, E. E.: J. Chem. Phys. 55, 2120 (1970).
152) Puckett, L. J., Teague, M. W.: J. Chem. Phys. 54, 2564 (1971).
153) Beggs, D. P., Field, F. H.: J. Chem. Soc. 93, 1567 (1971).
154) Field, F. H., Beggs, D. P.: J. Am. Chem. Soc. 93, 1576 (1971).
155) Puckett, L. J., Teague, M. W.: J. Chem. Phys. 54, 4860 (1971).
156) Zielinska, T. J., Wincel, H.: Chem. Phys. Letters 25, 354 (1974).
157) Davidson, W. R., Kebarle, P.: unpublished results, quoted in Ref. 134).
158) Tang, I. N., Castleman, A. W.: J. Chem. Phys. 57, 3638 (1972).
159) Tang, I. N., Castleman, A. W.: J. Chem. Phys. 60, 3981 (1974).
160) De Paz, M., Guidoni-Guardini, A., Friedman, L.: J. Chem. Phys. 52, 687 (1970).
161) Kraemer, W. P., Diercksen, G. H. F.: Theoret. Chim. Acta 27, 265 (1972).
162) Kistenmacher, H., Popkie, H., Clementi, E.: J. Chem. Phys. 61, 799 (1974).
163) O'Brien, E. F., Robinson, G. W.: J. Chem. Phys. 61, 1050 (1974).
164) Pottel, R.: Dielectric properties. In Ref. 29), Vol. 3, pp. 401—431.
165) Verall, R. E.: Infrared spectroscopy of aqueous electrolyte solutions. In Ref. 29), Vol. 3, pp. 211—264.
166) Lilley, T. H.: Raman spectroscopy of aqueous electrolyte solutions. In Ref. 29), Vol. 3, pp. 265—229.
167) Hertz, H. G.: Nuclear magnetic relaxation spectroscopy. In Ref. 29), Vol. 3, pp. 301—399.
168) Eigen, M., Maass, G.: Z. Physik. Chem. (neue Folge) 49, 163 (1966).
169) Diebler, H., Eigen, M., Ilgenfritz, G., Maass, G., Winkler, R.: Pure Appl. Chem. 20, 93 (1969).
170) Breitschwerdt, K. G.: Structural relaxation in water and ionic solutions. In: Structure of water and aqueous solutions, pp. 473—490 (Luck, W. A. P., ed.). Weinheim: Verlag Chemie 1974.
171) Fratiello, A., Lee, R. E., Nishida, V. M., Schuster, R. E.: J. Chem. Phys. 48, 3705 (1968).
172) Hertz, H. G., Raedle, C.: Ber. Bunsenges. Physik. Chem. 77, 521 (1973).
173) Bertagnolli, H.,Weidner, J., Zimmermann, H.W.: Ber. Bunsenges. Physik. Chem.78,2 (1974).
174) Narten, A. H.: J. Chem. Phys. 56, 5681 (1972).
175) Narten, A. H., Vaslow, F., Levy, H. A.: J. Chem. Phys. 58, 5017 (1973).
17c Falk, M., Knop, O.: Water in stoichiometric hydrates. In Ref.29), Vol. 2, pp. 55—113.
177) Eley, D. D., Evans, M. E.: Trans. Faraday Soc. 34, 1039 (1938).
178) Muirhead-Gould, J. S., Laidler, K. J.: Trans. Faraday. Soc. 63, 944 (1967).
179) Verwey, E. J.: Rec. Trav. Chim. 60, 887 (1941).
180) Eliezer, I., Krindel, P.: J. Chem. Phys. 57, 1884 (1972).
181) Morf, W. E., Simon, W.: Helv. Chim. Acta 54, 2683 (1971).
182) Tait, A. D., Hall, G. G.: Theoret. Chim. Acta 31, 311 (1973).
183) Heinzinger, K., Vogel, P.: Z. Naturforsch. 29a, 1164 (1974).
184) Sposito, G., Babcock, K. L.: J. Chem. Phys. 47, 153 (1967).
185) Hylton, J. E., Christofferson, R. E., Hall, G. G.: Chem. Phys. Letters 24, 501 (1974).
186) Yamabe, S., Kato, S., Fujimoto, H., Fukui, K.: Theoret. Chim. Acta 30, 327 (1973).
187) Klopman, G.: Chem. Phys. Letters 1, 200 (1967).
188) Germer, H. A.: Theoret. Chim. Acta. 34, 145 (1974).
189) Basu, S.: In: Advances in quantum chemistry (P. O. Löwdin, ed.), Vol. 1, p. 145. New York: Academic Press 1964.

190) Liptay, W.: In: Modern quantum chemistry, Istanbul lectures (O. Sinanoglu, ed.), Part II (1965), p. 173.
191) Chang, T. C., Habitz, P., Pittel, B., Schwarz, W. H. E.,: Theoret. Chim. Acta *34*, 263 (1974).
192) Kleiner, M., McWeeny, R.: Chem. Phys. Letters *19*, 476 (1973).
193) Bonifacic, V., Huzinaga, S.: J. Chem. Phys. *60*, 2779 (1974).
194) Marius, W.: Thesis Vienna (1974).
195) Breitschwerdt, K. G., Kistenmacher, H.: Chem. Phys. Letters *14*, 288 (1972).
196) Diercksen, G. H. F., Kraemer, W. P.: Theoret. Chim. Acta *23*, 387 (1972).
197) Pullman, A.: Intern. J. Quantum Chem., Symposium *1*, 33 (1974).
198) Kistenmacher, H., Popkie, H., Clementi, E.: J. Chem. Phys. *58*, 1689 (1973).
199) Pullman, A., Armbruster, A. M.: Intern. J. Quantum Chem., Symposium *8*, 169 (1974).
200) Port, G. N. J., Pullman, A.: Theoret. Chim. Acta *31*, 231 (1974).
201) Thorhallson, J., Fisk, C., Fraga, S.: Theoret. Chim. Acta *10*, 388 (1968).
202) Salzmann, J. J., Jørgensen, C. K.: Helv. Chim. Acta *51*, 1276 (1968).
203) Diercksen, G. H. F., Kraemer, W. P.: Chem. Phys. Letters *5*, 570 (1970).
204) Kistenmacher, H., Popkie, H., Clementi, E.: J. Chem. Phys. *58*, 5627 (1973).
205) Piela, L.: Chem. Phys. Letters *19*, 134 (1973).
206) Kistenmacher, H., Popkie, H., Clementi, E.: J. Chem. Phys. *59*, 5842 (1973).
207) Wigner, E.: Phys. Rev. *46*, 1002 (1934).
208) Diercksen, G. H. F., Kraemer, W. P.: Munich molecular program system-reference manual, special techn. report; Max Planck Institut für Physik und Astrophysik (to be published).
209) Merlet, P., Peyerimhoff, S. D., Buencker, R. S.: J. Am. Chem. Soc. *94*, 8301 (1972).
210) Delpuech, J. J., Serratrice, G., Strich, A., Veillard, A.: Chem. Commun. *1972*, 817.
211) Diercksen, G. H. F.: Theoret. Chim. Acta *21*, 335 (1971).
212) Rode, B. M., Preuss, H. W.: Theoret. Chim. Acta *35*, 369 (1974).
213) Rode, B. M.: Chem. Phys. Letters *26*, 350 (1974).
214) Ovchinnikov, Yu. A., Ivanov, V. T., Shkrob, A. M.: In: Molecular mechanisms of antibiotic action on protein biosynthesis and membranes (E. Munoz, F. Garcia-Fernandez and D. Vasquez, eds.), p. 459. Amsterdam: Elsevier 1972.
215) Gisin, B. F., Merrifield, R B.: J. Am. Chem. Soc. *94*, 6165 (1972).
216) Lentz, B. R., Scheraga, H. A.: J. Chem. Phys. *58*, 5298 (1973).
217) Diercksen, G. H. F., Kraemer, W. P.: Theoret. Chim. Acta *23*, 393 (1972).
218) Morf, W. E., Simon, W.: Helv. Chim. Acta *54*, 2689 (1971).
219) Kollman, P. A., Kuntz, I. D.: J. Am. Chem. Soc. *96*, 4766 (1974).
220) Burton, R. E., Daly, J.: Trans. Faraday Soc. *67*, 1219 (1971).
221) Burton, R. E., Daly, J.: Trans. Faraday Soc. *66*, 1281 (1970).
222) Lischka, H., Plesser, Th., Schuster, H.: Chem. Phys. Letters *6*, 263 (1970).
223) Russegger, P., Lischka, H., Schuster, P.: Theoret. Chim. Acta *24*, 191 (1972).
224) Russegger, P.: Thesis Vienna (1972).
225) Bunyatyan, B. Kh., Zhogolev, D. A., Ritschl, F.: Chem. Phys. Letters *24*, 520 (1974).
226) Zhogolev, D. A., Bunyatyan, B. Kh., Kruglyak, Yu. A.: Chem. Phys. Letters *18*, 135 (1973).
227) Kostetsky, P. V., Ivanov, V. T., Ovchinnikov, Yu. A., Shchembelov, G.: Febs Letters *30*, 205 (1973).
228) Schuster, P.: Chem. Phys. Letters *3*, 433 (1969).
229) Rode, B. M.: J. Chem. Soc. Faraday Trans. II *1973*, 1439.
230) Rode, B. M.: Chem. Phys. Letters *20*, 366 (1973).
231) Rode, B. M.: Chem. Phys. Letters *25*, 369 (1974).
232) Rode, B. M.: Communicated at the 10th Symposium on Theoretical Chemistry, Semmering, Austria (1974).
233) Schuster, P.: Theory of hydrogen bonding in water and ionic solutions. In: Structure of water and aqueous solutions (Proceedings of the International Symposium, Marburg 1973), p. 142 (W. A. P. Luck, ed.). Weinheim: Verlag Chemie 1974.
234) Cremaschi, P., Gamba, A., Simonetta, M.: Theoret. Chim. Acta *25*, 237 (1972).
235) Rode, B. M.: Monatsh. Chem. *105*, 308 (1974).

236) Hirschfelder, J. O. (ed.): Intermolecular forces. Advan. Chem. Phys., Vol. *12*, New York: Interscience 1967.
237) Stokes, R. H., Marsh, K. N.: Ann. Rev. Phys. Chem. *23*, 65 (1972).
238) Hadzi, D.: Vibrational spectroscopy of the hydrogen bond. In: Ref. 242), in press.
239) Luck, W. A. P.: Infrared studies of hydrogen bonding in pure liquids and solutions. In: Ref. 29), Vol. *2*, pp. 235—321.
240) Choppin, G. R., Downey, J. R.: Spectrochim. Acta *30A*, 43 (1974).
241) Del Bene, J. E.: Chem. Phys. Letters *24*, 203 (1974).
242) Schuster, P., Zundel, G., Sandorfy, C. (eds.): The hydrogen bond–recent developments in theory and experiment. Amsterdam: North-Holland (in press).
243) Covington, A. K., Jones, P. (eds.): Hydrogen bonded solvent systems. London: Taylor and Francis 1968.
244) Morokuma, K., Iwata, S., Lathan, W. A.: Molecular interactions in: (Daudel, R., Pullman, B., eds.) The world of quantum chemistry pp. 277-316. Dordrecht: D. Reidel 1974.
245) Sinanoğlu, O.: Theoret. Chim. Acta *33*, 279 (1974).
246) Sinanoğlu, O.: Three types of potential needed in predicting conformations of molecules in solution and their use. In: (Daudel, R., and Pullman, B., eds.) The world of quantum chemistry, pp. 265—276. Dordrecht: D. Reidel 1974.
247) Sinanoğlu, O.: Solvent effects on molecular associations. In: (Pullman, B., ed.) Molecular associations in biology, pp. 427—445. New York: Academic Press 1968.
248) Halicioğlu, T., Sinanoğlu, O.: Ann. N. Y. Acad. Sci. *158*, 308 (1968).
249) Diner, S., Malrieu, J. P., Claverie, P.: Theoret. Chim. Acta *13*, 1 (1969).
250) Malrieu, J. P., Claverie, P., Diner, S.: Theoret. Chim. Acta *13*, 18 (1969).
251) Diner, S., Malrieu, J. P., Jordan, F., Gilbert, M., Theoret. Chim. Acta *15*, 100 (1969).
252) Pullman, B., Port, J.: Mol. Pharmacol. *10*, 360 (1974).
253) Pullman, B., Courriere, P., Berthod, M.: J. Medic. Chem. *17*, 439 (1974).
254) Diercksen, G. H. F., Kraemer, W. P., von Niessen, W.: Theoret. Chim. Acta *28*, 67 (1972).
255) Kollman, P. A., Allen, L. C.: J. Am. Chem. Soc. *93*, 4991 (1971).
256) Del Bene, J., Pople, J. A.: J. Chem. Phys. *52*, 4858 (1970).
257) Morokuma, K., Winick, J.: J. Chem. Phys. *52*, 1301 (1970).
258) Ditchfield, R., Hehre, W. J., Pople, J. A.: J. Chem. Phys. *54*, 724 (1971).
259) Whitten, J. L.: J. Chem. Phys. *56*, 5458 (1972).
260) Bell, R. P.: Advan. Phys. Org. Chem. *4*, 1 (1966).
261) Bell, R. P., Clunie, J. C.: Trans. Faraday Soc. *48*, 439 (1952).
262) Del Bene, J. E.: J. Chem. Phys. *58*, 3139 (1973).
263) Schuster, P., Jakubetz, W., Beier, G., Meyer, W., Rode, B. M.: Potential curves for proton motion along hydrogen bonds. In: (Bergmann, E. D. and Pullman, B., eds.), Chemical and biochemical reactivity. Jerusalem: Academic Press, in press.
264) Ilse, F. E., Hartmann, H.: Z. Physik. Chem. *197*, 239 (1951).
265) Ilse, F. E., Hartmann, H.: Z. Naturforsch. *6a*, 751 (1951).
266) Dahl, J. P., Ballhausen, C. J.: Molecular orbital theories of inorganic complexes (P. O. Löwdin, ed.). Advan. Quantum Chem. *4*, 170 (1968).
267) Slater, J. C., Statistical exchange-correlation in the self-consistent field (P. O. Löwdin, ed.). Advan. Quantum Chem. *6*, 1 (1972).
268) Johnson, K. H.: Scattered-wave theory of the chemical bond (P. O. Löwdin, ed.). Advan. Quantum Chem. *7*, 143 (1973).
269) Dreyfus, M., Pullman, A.: Theoret. Chim. Acta *19*, 20 (1974).
270) Gupta, A., Rao, C. N. R.: J. Chem. Phys. *77*, 2888 (1973).
271) Balasubramanian, D., Goel, A., Rao, C. N. R.: Chem. Phys. Letters *17*, 482 (1972).
272) Cremaschi, P., Simonetta, M.: Theoret. Chim. Acta *37*, 341 (1975).
273) Chan, R. E., Clementi, E., Kress, J. W., Lie, G. C.: J. Chem. Phys. *62*, 2026 (1975).
274) Watts, R. O., Clementi, E., Fromm, J.: J. Chem. Phys. *61*, 2550 (1975).
275) Fromm, J., Clementi, E., Watts, R. O.: J. Chem. Phys. *62*, 1388 (1975).
276) Vogel, P. C., Heinzinger, K.: Z. Naturforsch. *30a*, 789 (1975).

Received November 14, 1974

Conjectures on the Structure of Amorphous Solid and Liquid Water

Dr. Stuart A. Rice

Department of Chemistry and The James Franck Institute, University of Chicago, Chicago, Illinois, 60637, U.S.A.

Contents

I. Introduction

Another article on water? Why?

Anyone surveying the huge literature dealing with experimental and theoretical studies of water is bound to ask these questions.

This article describes some conjectures concerning the structures of amorphous solid and liquid water, and the considerations underlying those conjectures. The approach adopted originally derived from a negative reaction to the status of the available information. Enormous effort has been invested in the experimental determination of the properties of water and water based solutions, in attempts to interpret those properties in terms of molecular interactions, and in the development of models with which known properties can be correlated and unknown properties predicted [1]. Despite the effort expended our factual knowledge is meagre and our understanding rudimentary. For, under ordinary conditions the interpretation of the observed properties of water is rendered difficult by the simultaneous presence of positional, orientational and thermal disorder, along with a small but not negligible molecular dissociation. Furthermore, neither the existing theories of the liquid state [2], nor the available methodologies of statistical mechanics have yet provided a compact, useful, description of liquids composed of molecules between which there are strong non-central, saturable, forces, such as exist in water. Consequently, excepting the elegant, important, computer generated simulations of a model of water by Rahman and Stillinger [3], theory has not guided experiment, nor experiment aided in the selection of useful theoretical concepts and/or approximations.

The preceding observations stimulated Olander and Rice [4] to search for a substance that is simultaneously "simpler" than water yet a "good model" of it. They suggested that amorphous solid water [$H_2O(as)$], first reported by Burton and Oliver [5] in 1935, satisfied these two requirements. Unlike the liquid, amorphous solid water can be studied at low temperature where the effects of thermal excitation and positional and orientational disorder can be separated. Moreover, it is plausible to accept as a working hypothesis that the amorphous solid is, essentially, extensively supercooled liquid water; if so, the properties of the amorphous solid should be directly related to those of the liquid.

This article summarizes the results of some recent studies of amorphous solid water [6-8], suggests a correlation of its properties with those of the liquid, and drawing on the common features of a variety of models, projects a conjecture concerning the structure of water. The structure proposed is not entirely new; given the range of speculation in the literature that is not surprising. What is different is that new information about $H_2O(as)$ permits the construction of a testable hypothesis concerning its structure, and by extrapolation also that of the liquid. I hope to convince the reader through this report of research in progress that further studies of $H_2O(as)$ will be rewarding, particularly with respect to providing insights into the structure of the liquid.

Note added in proof: Since completion of this manuscript in December 1974 much new information has become available. Wherever possible this information has been integrated into the text via notes added in proof.

II. Preparation of Amorphous Solid Water

At a given (low) temperature and pressure a crystalline phase of some substance is thermodynamically stable *vis a vis* the corresponding amorphous solid. Furthermore, because of its inherent metastability, the properties of the amorphous solid depend, to some extent, on the method by which it is prepared. Just as in the cases of other substances, H_2O(as) is prepared by deposition of vapor on a cold substrate. In general, the temperature of the substrate must be far below the ordinary freezing point and below any possible amorphous → crystal transition point. In addition, conditions for deposition must be such that the heat of condensation is removed rapidly enough that local crystallization of the deposited material is prevented. Under practical conditions this means that, since the thermal conductivity of an amorphous solid is small at low temperature, the rate of deposition must be small.

In most of the early studies [9] of H_2O(as) the vapor was condensed on metal surfaces in the temperature range $77 \, K < T < 153 \, K$. Immediately above 153 K cubic ice Ic is always the dominant component in the deposit. Examination of the available diffraction data, supplemented by new experimental studies, convinced Olander and Rice that most deposits obtained at or above 77 K are likely contaminated with crystalline ice. They established conditions for the deposition of pure H_2O(as) on a variety of substrates [10]. Briefly put, the temperature of the substrate should be low, preferably below 55 K, and the rate of deposition very small (a few mg/hour). There is evidence that H_2O(as) can be deposited on a substrate at 77 K if the deposition is slow enough. The use of high deposition rates at 77 K leads to polycrystalline ice Ic mixed with H_2O(as). A sample of pure H_2O(as) is stable "indefinitely long" (at least several months) if maintained below ~20 K. At about 135 K, with some variation from sample to sample, the amorphous solid transforms spontaneously and irreversibly to ice Ic.

The procedure described for the preparation of H_2O(as) has one serious drawback, already mentioned, which is simultaneously an interesting and promising potential benefit. It is well known that the phase diagram of ice has regions of stability for many different crystalline modifications [11], thereby testifying to the existence of a complex multiminimum many molecule potential energy surface. Given that H_2O(as) is a metastable substance, and that its formation depends on the kinetics of the deposition process, it is possible (likely?) that materials prepared under different conditions can have different structures. Indeed, there is some evidence that H_2O(as) prepared at ~10 K is different from H_2O(as) prepared at ~77 K (see next section).

Note added in proof: It is only very recently that it has been recognized that there can be several forms of H_2O(as). Consequently, not all of the relevant deposition parameters have been adequately controlled in each experiment. Even when pure H_2O(as) has been made, the form deposited may not be that expected from only a knowledge of the temperature of the substrate. In future experimental studies more attention will have to be given to, *e.g.* the angle of incidence of the vapor on the substrate, the rate of deposition and the consequent surface temperature, etc.

III. X-Ray and Neutron Diffraction Studies

The structure of a liquid is conventionally described by the set of distributions of relative separations of atom pairs, atom triplets, etc. The fundamental basis for X-ray and neutron diffraction studies of liquids is the observation that in the absence of multiple scattering the diffraction pattern is completely determined by the pair distribution function.

For the case of a pure monatomic liquid, in the limit that there are only pair interactions, the pair distribution function provides a complete microscopic specification from which all thermodynamic properties can be calculated [2]. If there are (excess) three molecule interactions, then one must also know the triplet distribution function to complete the microscopic description; the extension to still higher order (excess) interactions is obvious.

The case of a polyatomic liquid is more complicated by virtue of the presence of more than one type of atom and the presence of coexisting intramolecular and intermolecular structures. Two forms of description are commonly used. When the temperature is low enough that the molecule can be considered a rigid body the relative configuration of a pair of molecules is specified by the vector separation of the centers of mass and the three Euler angles defining the orientation of one with respect to the other. The molecular pair distribution function obviously depends on the same set of variables. However, given that the liquid is macroscopically isotropic it is not possible to obtain the full molecular pair distribution function from one dimensional (intensity versus scattering angle) diffraction data. It *is* possible to represent the molecular pair distribution function as a series expansion in the complete set of orthonormal polynomials in the orientation angles, with coefficients that depend only on the separation of the centers of mass of the molecules [12]. When applied to the case of scattering from a collection of anisotropic molecules, it is found that the leading terms in the scattered intensity are dependent on the spherically symmetric contributions to the pair distribution function, and that the higher order terms describe the influence on the scattered intensity of molecular orientation. For molecules that are only slightly asymmetric only the leading terms in the series referred to need be retained [13], and the intensity of coherently scattered radiation per molecule takes the form [14]

$$I(s) = \langle F^2 \rangle + \langle F \rangle^2 \int\limits_0^\infty dR \ 4\pi R^2 \ \varrho \, h_{00}(R) \frac{\sin sR}{sR} \tag{3.1}$$

where $\langle F^2 \rangle$ is the total form factor for scattering by an isolated molecule, $\langle F \rangle^2$ the spherical (orientationally averaged) part of $\langle F^2 \rangle$, $s \equiv \dfrac{4\pi}{\lambda}\sin\theta$ with λ the wavelength and 2θ the scattering angle, and $h_{00}(R) \equiv g_{00}(R) - 1$ is the spherical part of the pair correlation function for molecular centers.

Given that X-rays are scattered from the electrons of the molecule and that the electron density contours of the water molecule deviate very little from spherical symmetry, it is not surprising that the X-ray form factors $\langle F^2 \rangle_{\text{X-ray}}$ and $\langle F \rangle^2_{\text{X-ray}}$ differ negligibly ($\sim 1\%$) [1,13]. Thus, the water molecule appears spherically symmetric to X-rays, and orientational correlation between water molecules in the liquid cannot be detected by X-ray scattering. However, X-ray diffraction data can be used to extract the spherical part of the molecular pair distribution function.

In contrast with the scattering of X-rays by electrons, it is the nuclei that scatter neutrons. It is to be expected, then, that neutron scattering from a molecule is far from isotropic, and that many terms in the orthonormal polynomial expansion must be kept to provide an accurate representation of the molecular pair distribution function. Accordingly, a different representation of the scattered intensity is more convenient. Recalling that in the limit that multiple scattering is negligibly small the intensity of coherent scattering depends only on the distribution of pairs of scattering centers, one can write

$$I(s) = \sum_{\alpha=1}^{n}{}' f_\alpha^2 + \sum_{\alpha=1}^{n}{}' \sum_{\beta \neq \alpha=1}^{n}{}' f_\alpha f_\beta (\varrho_\alpha \varrho_\beta)^{1/2} \int_0^\infty dR_{\alpha\beta} 4\pi R_{\alpha\beta}^2 h_{\alpha\beta}(R_{\alpha\beta}) \frac{\sin s R_{\alpha\beta}}{s R_{\alpha\beta}} \qquad (3.2)$$

with f_α the atomic coherent scattering amplitude, ϱ_α the bulk density of atoms of type α, $h_{\alpha\beta} \equiv g_{\alpha\beta} - 1$ is the atomic $\alpha\beta$ pair correlation function, and the summation is over the n atoms in one molecule. Clearly, to derive the three atom pair correlation functions for water, $h_{OO}(R)$, $h_{OH}(R)$ and $h_{HH}(R)$, from diffraction data requires three independent experiments on water isomers which have (presumably) the same structure but different $\{f_\alpha\}$. In practice only two such experiments are possible, namely X-ray scattering from H_2O or D_2O and neutron scattering from D_2O. Thus, at least for the present, it is not possible to determine from experimental data only all three of the distributions $h_{OO}(R)$, $h_{OH}(R)$ and $h_{HH}(R)$. Nevertheless, the available experimental data can be combined with considerations based on models to yield valuable information about the structure of water [15].

Note added in proof: Recent work by E. Kalman and G. Palinkas (Proceedings of Conference on Liquid Structure Studies, Keszthely, Hungary, August 1974) suggests that electron diffraction from water can be a useful adjunct to other studies. This follows from the observation that the ratio of oxygen to hydrogen scattering factors, as a function of scattering angle, is different enough for X-rays, electrons and neutrons, that the OO, OH and HH distributions contribute differently to the overall scattering pattern.

A. Liquid Water

X-ray diffraction from water has been studied intermittantly since 1930. All the available reports are in satisfactory agreement; we choose to discuss the extensive high quality data from the Oak Ridge National Laboratory [16]. Figure 1 shows the structure function

$$\tilde{h}_{OO}(s) = \varrho \int_0^\infty dR\, 4\pi R^2 h_{OO}(R) \frac{\sin s R}{s R} \qquad (3.3)$$

and Fig. 2 the pair correlation function $h_{OO}(R)$ descriptive of molecular centers (essentially the OO correlation function) for liquid water for several tempera-

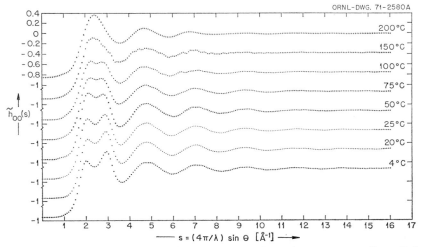

Fig. 1. The structure functions for liquid water at a variety of temperatures (from Ref. [16])

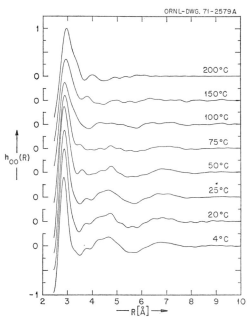

Fig. 2. OO pair correlation functions for liquid water at a variety of temperatures (from Ref. [16])

tures, taken from the work of Narten and Levy. The most important features of $\tilde{h}_{OO}(s)$ are the double maximum (below 100 °C), and the oscillations in amplitude about the asymptotic value $\tilde{h}_{OO}(\infty) = 0$. Because of the inverse relationship between the scattering variable s and the pair separation coordinate R, the region $s \lesssim 4$ Å^{-1} reflects the influence of interactions between nearest neighbor molecules, and the region $s < 4$ Å^{-1} that of molecules further apart. The frequency, amplitude and rate of damping of the oscillations about $\tilde{h}_{OO}(\infty) = 0$ are related, respectively, to the position, area and width of the nearest neighbor peak of $h_{OO}(R)$. This peak indicates a nearest neighbor separation of 2.84 Å at 4 °C increasing to 2.94 Å at 200 °C, and the area corresponds to approximately four nearest neighbors, the latter invariant with respect to temperature. The clear implication of these results is that the local environment of a water molecule is tetrahedral, at least on average. This interpretation is supported by the measured ratio (1.63) of distances corresponding to the first and second peaks of $h_{OO}(R)$, which also is in accord with local tetrahedral coordination of the water molecule. The tetrahedral coordination is, however, not perfect, since the best estimate of the number of nearest neighbors exceeds four ($\sim$4.4), and the breadth of the second neighbor peak "centered" at 4.5 Å indicates sizable deviations from a simple regularly propagated tetrahedral geometry.

As the temperature is increased the double peak structure of $\tilde{h}_{OO}(s)$ becomes less and less pronounced, and by 150 °C there is only a single peak near $s \approx 2.5$ Å^{-1}. In direct (R) space, this change corresponds to the loss of correlation between molecular centers in the region $R \lesssim 4$ Å, shown by the decrease in amplitude of the oscillations of $h_{OO}(R)$ as T increases. Despite this dramatic change for $R \lesssim 4$ Å, local tetrahedrality remains the dominant feature of the structure of the liquid.

For very large R the behavior of $h_{OO}(R)$ is determined by long wavelength density fluctuations, hence by the compressibility of the liquid. Precision small angle X-ray scattering measurements [17], as well as extrapolations of $\tilde{h}_{OO}(s)$ to $s = 0$ for the data shown, are in agreement with predictions based on bulk thermodynamic data.

The structure functions determined from neutron scattering by D_2O, from the measurements of Page and Powles [18] and from the measurements of Narten [19], are not in complete agreement (see Fig. 3), the difference being most pronounced

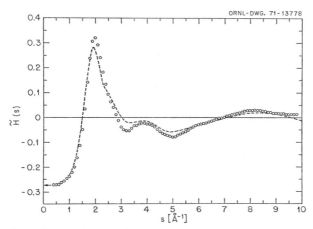

Fig. 3. A comparison of neutron scattering data from the experiments of Narten (o) and of Powles (---) (from Ref. [19])

in the range $2\,\text{Å}^{-1} < s < 6\,\text{Å}^{-1}$. Narten discusses this discrepancy; we refer the reader to his paper [19] for details. Because of the extent of consistency with X-ray data and model considerations we choose to describe Narten's results.

The neutron structure function for water is of the form

$$
\tilde{H}(s) \equiv \left[I(s) - \sum_{\alpha=1}^{3} f_{\alpha}^{2} \right] \left[\sum_{\alpha=1}^{3} f_{\alpha} \right]^{-2}
$$

$$
= \frac{\left[\sum_{\alpha=1}^{3} \sum_{\beta=1}^{3} f_{\alpha} f_{\beta} \exp\left(-b_{\alpha\beta} s^{2}\right) \dfrac{\sin s R_{\alpha\beta}}{s R_{\alpha\beta}} \right] - \sum_{\alpha=1}^{3} f_{\alpha}^{2}}{\left[\sum_{\alpha=1}^{3} f_{\alpha} \right]^{2}}
$$

$$
+ \sum_{\alpha=1}^{3} \sum_{\beta \neq \alpha=1}^{3} f_{\alpha} f_{\beta} \tilde{h}_{\alpha\beta}(s) \left[\sum_{\alpha=1}^{\alpha} f_{\alpha} \right]^{-2}
$$

$$
= \tilde{h}_{1}(s) + 0{,}09\,\tilde{h}_{\text{OO}}(s) + 0.42\,\tilde{h}_{\text{OD}}(s) + 0.49\,\tilde{h}_{\text{DD}}(s)
$$

$$(3.4)$$

with $b_{\alpha\beta}$ the mean square variation of the distance $R_{\alpha\beta}$. Note that $\tilde{H}(s)$ is almost completely dominated by $\tilde{h}_{\text{OD}}(s)$ and $\tilde{h}_{\text{DD}}(s)$, the contribution from the oxygen-oxygen correlation function amounting to only 9% of the total scattering. Thus the neutron structure function is much more sensitive to orientational correlation between pairs of molecules than to their positional correlation, thereby nicely complementing the X-ray structure function. Narten takes $\tilde{h}_{\text{OO}}(s)$ from his X-ray studies and subtracts its contribution to the neutron structure function, generating thereby a weighted sum of $\tilde{h}_{\text{OD}}(s)$ and $\tilde{h}_{\text{DD}}(s)$. The result is shown in Fig. 4 Further separation of the data into unique functions $\tilde{h}_{\text{OD}}(s)$ and $\tilde{h}_{\text{DD}}(s)$ requires another independent diffraction experiment not now available.

The interpretation of the correlation function derived from neutron diffraction in the range $R < 2.8\,\text{Å}$ is straightforward. The first peak, at $0.94\,\text{Å}$, arises from the intramolecular OD spacing, that at $1.7\,\text{Å}$ is likely a superposition of the intramolecular DD and the intermolecular hydrogen bonded OD separations, those at $2.4\,\text{Å}$ and $3.4\,\text{Å}$ are probably OD and DD separations between near neighbor molecules.

Lengyel and Kalman have reported an electron diffraction study of water [20]. The high energy electrons used permit the structure function to be probed at (large) values of s unattainable in either X-ray or neutron diffraction. This feature is valuable in that the wider the range of s for which data are available, the more accurate is the inversion of the observed $\tilde{h}(s)$ to the direct space function $h(R)$. Typical data, for D_2O at 25 °C, are shown in Fig. 5. Note the similarities, and differences, between the electron diffraction, X-ray diffraction and neutron diffraction data.

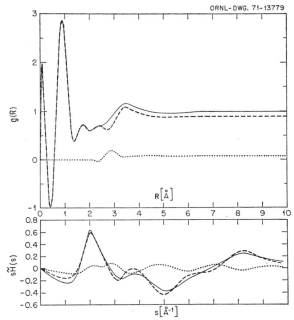

Fig. 4. Weighted structure function

$$s\tilde{H}(s) = \tilde{h}_1(s) + \tilde{H}_2(s)\,\tilde{h}_{OO}(s)$$

$$\tilde{H}_2(s) = \left[f_0 + 2f_D \exp\left(-b_{OD}s^2\right) \frac{\sin sR_{OD}}{sR_{OD}} \right]^2 [f_0 + 2f_D]^{-2}$$

for liquid water (D_2O) at 25°C. . . . . . CO contribution, – – – difference curves descriptive of OD and DD correlations. (from Ref. [19])

B. Amorphous Solid Water

Just as for liquids, the most convenient representation of the structure of an amorphous solid is through the directly measurable pair distribution function. Wenzel, Linderstrøm-Lang and Rice [6] have studied the coherent scattering of neutrons from D_2O(as) at 10 K. As shown in Fig. 6, the overall radial distribution function has well defined peaks at 1.00, 1.68 and 2.26 Å; the widths of these peaks are determined by the limited resolution of the experiment. Note the similarities between these peak distances and the corresponding ones in the neutron diffraction spectrum of the liquid. Clearly, the interpretation is the same. The first peak is due to the OD separation within a water molecule. The value quoted for this separation is, within the experimental error, the same as the corresponding distances in other hydrogen bonded water structures [21]. The second peak is likely a superposition of the intramolecular DD and intermolecular hydrogen bonded O...D separations, and the third peak is likely the nearest neighbor intermolecular DD separation. These data imply that the molecule in D_2O(as) has the same dimensions as in other condensed phases of water, and that there is hydrogen bonding and tetrahedral coordination. Once again the use of the term "tetrahedral coordination" does not imply that all angles are exactly 109.5°. Rather it is intended to describe the qualitative characteristics of the local environment. In general, we expect there to be a range of angles, centered about the tetrahedral value, consistent with both four coordination and high density.

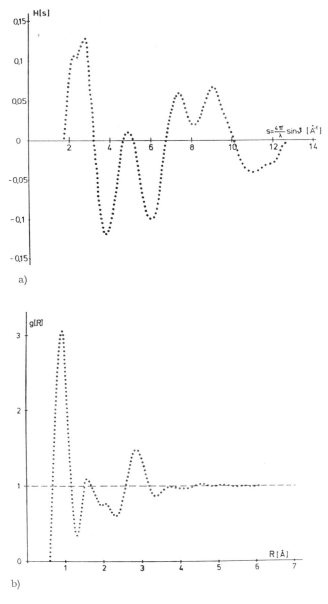

a)

b)

Fig. 5a and b. The structure function (a) and pair correlation function (b) deduced from electron diffraction studies of D_2O (Kalman and Palinkas, private communication)

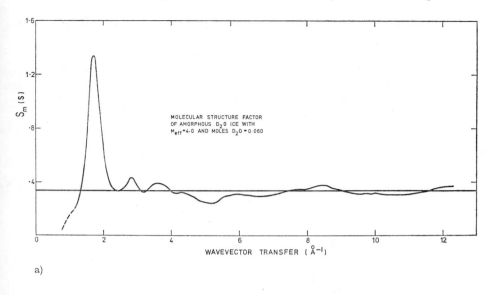

a)

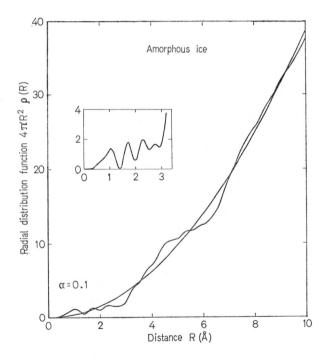

b)

Fig. 6a and b. The structure function (a) and radial distribution function (b) of D$_2$O(as) (from Ref. [6]). The inset refers to a different convergence parameter α in the inversion of the structure function

An example of what can be expected from this kind of structural disorder is provided by amorphous Ge.

In the case of Ge(as) [22-24], the bond angles have a root mean square deviation of ~7° about 109°, and the bond length differs insignificantly (rms deviation of ~1%) from that in crystalline Ge. Clearly, there can be two sources of bond angle variation in H_2O(as), namely deformation of the intramolecular HOH angle and deviation of the H...OH bond from the near linearity characteristic of ice Ih. It seems very likely that the latter is much more important than the former. In fact, the two possibilities can be distinguished. The distribution of intramolecular DOD angles in the amorphous solid can be measured by neutron scattering if the range of scattering angle can be sufficiently extended, since at very large s the molecular form factor dominates the s dependence of the scattering. This difficult experiment has been undertaken by J. Wenzel [25].

As noted earlier, the diffraction of X-rays, unlike the diffraction of neutrons, is primarily sensitive to the distribution of OO separations. Although many of the early studies [9] of amorphous solid water included electron or X-ray diffraction measurements, the nature of the samples prepared and the restricted angular range of the measurements reported combine to prevent extraction of detailed structural information. The most complete of the early X-ray studies is by Bondot [26]. Only scanty description is given of the conditions of deposition but it appears likely his sample of amorphous solid water had little or no contamination with crystalline ice. He found a liquid-like distribution of OO separations at 83 K, with the first neighbor peak centered at 2.77 Å. If the pair correlation function is decomposed into a superposition of Gaussian peaks, the area of the near neighbor peak is found to correspond to 4.23 molecules, and to have a root mean square width of 0.50 Å.

A much more detailed X-ray diffraction study has been carried out by Venkatesh, Rice and Narten, who prepared samples at both 10 K [7] and 77 K [27], and obtained diffraction data at these temperatures.

For comparison purposes they also studied the diffraction pattern of polycrystalline ice Ih at 77 K. We shall discuss these data in detail, drawing heavily on Ref. [27]. It is convenient to begin with the case of polycrystalline ice Ih, and discuss in turn H_2O(as) deposited at 77 K, H_2O(as) deposited at 10 K, and the comparison with the neutron diffraction data.

C. Polycrystalline Ice Ih at 77 K

In the structure function for polycrystalline ice Ih (Fig. 7a) the (100), (002), and (101) reflections are not resolved but appear as a single peak centered around 1.75 Å^{-1}. Most of the higher-order reflections are well resolved at positions in excellent agreement with those calculated from the cell parameters. Thermal broadening of the Bragg peaks increases rapidly with increasing momentum transfer, and for values of $s > 12$ Å^{-1} the structure function shows only the expected sinusoidal oscillations from nearest-neighbor oxygen atom pairs.

The oxygen atom pair correlation function for crystalline ice Ih is shown in Fig. 7b. Note that the first peak in the function $h_{OO}(R)$ is completely resolved. The NVR analysis of the large s part of the structure function indicates that this

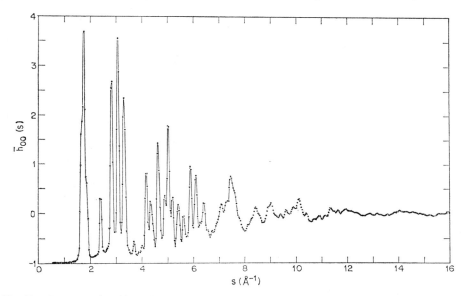

Fig. 7a. Structure function of polycrystalline ice Ih prepared by vapor deposition at 77 K from X-ray diffraction (from Ref. [27])

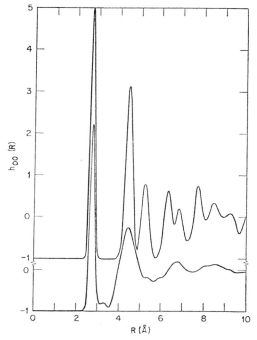

Fig. 7b. Oxygen atom pair correlation functions for polycrystalline ice Ih (top) and for H_2O(as) (bottom), both prepared and studied at 77 K, derived from X-ray diffraction (from Ref. [27])

near-neighbor peak is quantitatively described by a Gaussian distribution of distances with mean value $R_{OO} = 2.751(2)$ Å and rms deviation $l_{OO} = 0.091(3)$ Å. The distance compares well with the result from neutron diffraction studies on single crystals of (D_2O) ice Ih at 123 K, namely $R_{OO} = 2.75$ Å[1].

NVR have analyzed the entire structure function of crystalline ice Ih in terms of Gaussian distance distributions. The mean oxygen-oxygen separations, to a radial distance of 10 Å, and the number of pairs in this range, were generated from the cell parameters and atomic positions in the unit cell. The quantities to be found by least-squares refinement were the rms variations in interatomic distance which cannot be determined from the crystal data. NVR find that the rms-variation, l_2, in the second neighbor distance, R_2, adjusted by least squares, is sufficient to determine the rms-variations in the longer distances through the relation $l_{\alpha\beta}^2/R_{\alpha\beta} = l_2^2/R_2$. Correlation functions obtained by Fourier inversion of the computed structure functions were found to be in quantitative agreement with the functions $h_{OO}(R)$ derived from the diffraction data.

Since the mean OOO angle $\theta = 109.5°$ in ice Ih is determined by the distance ratio of the peaks descriptive of first and second neighbor distributions, one can estimate its rms-variation, $\langle(\Delta\theta)^2\rangle^{\frac{1}{2}}$. If it is assumed that the variance in this angle is given by the variance in the second neighbor distance alone, its magnitude should be $\langle(\Delta\theta)^2\rangle^{\frac{1}{2}} = 0.166/2.751 = 0.06$ radians, or about $3°$.

Finally, the density of crystalline ice Ih derived by NVR is 0.93 gm cm^{-3}. Given the uncertainty in the determination, which depends on a fitting of the structure function for small R, this can be considered identical to the density of 0.924 gm cm^{-3} computed from the cell parameters.

D. H_2O(as) 77 K/77 K

Correlation functions $h_{OO}(R)$ for the amorphous deposit prepared and studied at 77 K are shown in Fig. 7b together with the curve for polycrystalline ice Ih. As in the crystalline phase, the nearest-neighbor oxygen-oxygen correlations in H_2O(as) occur in an exceptionally narrow band centered at 2.76 Å, with rms-deviation 0.114 Å. The distance ratio for second and first neighbors indicates tetrahedral coordination on the average, but the second neighbor peak near 4.5 Å is much broader than in crystalline ice. The two major peaks at 2.76 Å and 4.5 Å are not completely separated; for the present we do not ascribe structural significance to the small feature near 3.3 Å.

Just as for crystalline ice Ih, the structure function for H_2O(as) 77 K/77 K can be well fitted with three adjustable parameters: the nearest neighbor OO distance, its rms-variation, and the variance in the second neighbot distance. All other distances are generated by the symmetry of the ice Ih structure and the rms-variations in longer distances from the relation $l_{\alpha\beta}^2/R_{\alpha\beta} = l_2^2/R_2$. From the variance in the second neighbor distance distribution NVR estimate a rms-variation of $8°$ in the local OOO angle about the mean tetrahedral value. The agreement of the structure and correlation functions for the described model with the curves derived from X-ray diffraction is not unique and hence not sufficient proof of its reality.

The density of H_2O(as) **77 K/77 K** derived from the X-ray data is 0.94 gm cm^{-3} in good agreement with the value 0.94 (2) gm cm^{-3} obtained for H_2O(as) by a more direct technique (see Section VI).

E. H_2O(as) 10 K/10 K

The correlation function $h_{OO}(R)$ for the amorphous deposit prepared and studied at 10 K is shown in Fig. 7c together with the curve for H_2O(as) 77 K/77 K. As in H_2O(as) **77 K/77 K**, the nearest-neighbor oxygen-oxygen correlations in H_2O(as) **10 K/10 K** occur in a narrow band centered at 2.76 Å. The rms-variation in this distance, 0.153 Å, is significantly larger than the corresponding value for H_2O(as) **77 K/77 K**, namely 0.114 Å, indicating a wider distribution of local OO distances about the mean value of 2.76 Å in the low-temperature deposit. The distance ratio for second and first neighbors indicates tetrahedral coordination on the average, and the variance in the second neighbor distance distribution indicates a local variation of about 8° about the man tetrahedral value.

A striking difference between the function $h_{OO}(R)$ for H_2O(as) **10 K/10 K** and for H_2O(as) **77 K/77 K** is apparent in the region of radial distances 3 Å $< R <$ 4 Å: The curve for low-temperature H_2O(as) shows a partially resolved peak at

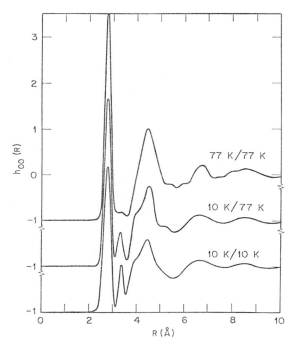

Fig. 7c. Oxygen atom pair correlation functions for H_2O(as) derived from X-ray diffraction. Top curve is for **77 K** deposit studied at **77 K**; bottom curves are for **10 K** deposit measured at **10 K** and at **77 K**. (From Ref. [27])

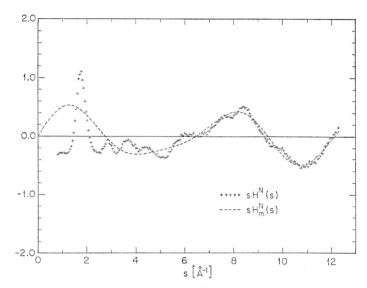

Fig. 7d. Weighted neutron structure function, $s\tilde{H}^N(s)$, derived from data of Wenzel, Linderstrom-Lang and Rice (Ref. [6]). The molecular structure function is $s\tilde{H}_m^N(s)$, and is calculated (from Ref. [27])

3.3 Å and a shoulder near 4 Å. The area under this peak or, more precisely, the integral

$$N = 4 \pi \varrho \int_{R_1}^{R_2} R^2 g_{OO}(R)\, dR \qquad (3.5)$$

with $R_1 = 3.1$ Å and $R_2 = 3.6$ Å yields a value of 1.4 water molecules. Another significant difference between high- and low-temperature $H_2O(as)$ is the density of 1.1 (1) gm cm^{-3} for the 10 K deposit, estimated from the X-ray data.

The ice I model used to describe the structure of $H_2O(as)$ 77 K/77 K does not predict any oxygen-oxygen distances in the region 2.76 Å $< R <$ 4.5 Å and is hence not capable of explaining the observed distance spectrum and density of $H_2O(as)$ 10 K/10 K. We defer to a later section the discussion of an appropriate model for this high density $H_2O(as)$.

In summary, the X-ray data show that positional correlation between oxygen atoms, i.e., structure, in $H_2O(as)$ extends to about 15 Å, about twice as far as in liquid water near the melting point. As in crystalline ice Ih, the nearest-neighbor correlations in $H_2O(as)$ occur in a narrow band centered at 2.76 Å. Unlike crystalline ice Ih, the distribution of second neighbors at 4.5 Å in $H_2O(as)$ is very broad and the variance in this distance distribution suggests a rms-variation in local OOO angles of $\sim 8°$ about the mean tetrahedral value. An angular spread of this magnitude is sufficient to explain the loss of all positional correlation a few molecular radii away from any starting point.

The radial distribution of oxygen atoms in $H_2O(as)$ prepared at 10 K is qualitatively different from that in $H_2O(as)$ prepared at 77 K. The high-temperature phase has the same density as crystalline ice I, and the average structure of this material can be described as a continuous random network of a single parentage, perhaps of the ice I type. The low-temperature phase has a density of 1.1 gm cm^{-3}, and the average structure of this material cannot be described in terms of a random network model of a single parentage.

F. Comparison with Neutron Diffraction Data

In the neutron diffraction experiments of Wenzel, Linderstrom-Lang, and Rice (WLR) [6] the sample was prepared by vapor deposition at a rate of ~10 mg/hour on a cadmium substrate maintained at a temperature of about 7 K.

At the time the neutron diffraction experiments were carried out it was not known that there are two forms of amorphous solid water. Consequently, although the deposition system was designed to ensure elimination of crystalline ice in the sample, neither the geometry nor the deposition rate were the same as used in the X-ray experiments of Narten, Venkatesh and Rice [7,27]. We shall argue below that although the substrate temperature used by WLR was low, their data are only consistent with diffraction from high temperature low density $D_2O(as)$.

The total neutron structure function may be written as the sum of two terms, namely

$$\tilde{H}^N(s) = \tilde{H}_m^N(s) + \tilde{H}_d^N(s) . \tag{3.6}$$

The first term, $\tilde{H}_m^N(s)$, describes the structure of D_2O molecules and the second term, $\tilde{H}_d^N(s)$, contains information about orientational and distance correlations between pairs of molecules. The function $\tilde{H}_d^N(s)$ decays to zero value much faster than the function $\tilde{H}_m^N(s)$, so that at large values of the variable s the total structure function is almost completely determined by the molecular structure. This fact may be used to determine the mean distances, $R_{\alpha\beta}$, and their rms-variations, $l_{\alpha\beta}$, by least-squares refinement of the molecular structure function

$$\tilde{H}_m^N(s) = [\sum_\alpha |f_\alpha|]^{-2} \sum_\alpha \sum_{\alpha \neq \beta} e^{-l_{\alpha\beta}^2 s^2/2} \frac{\sin s R_{\alpha\beta}^*}{s R_{\alpha\beta}} \tag{3.7}$$

against the large s part of the total structure function derived from the neutron data. The quantity $R_{\alpha\beta}^*$ is related to the average distance $R_{\alpha\beta}$ by the expression

$$R_{\alpha\beta}^* = R_{\alpha\beta} - \frac{l_{\alpha\beta}^2}{R_{\alpha\beta}} \tag{3.8}$$

Eq. (3.8) assumes harmonic vibrational motion, a satisfactory approximation for the range of momentum transfers covered by the neutron data. NVR adjusted the OD distance, its rms-variation, and the rms-variation in the intramolecular DD distance by least-squares. The DD distance itself was computed from the OD distance and an assumed DOD angle of 104.5°, because its value, 1.58 Å, is too close to the hydrogen-bonded $D_2O...D$ distance expected near 1.8 Å. The molecular structure function $\tilde{H}_m^N(s)$ calculated from these parameters is compared with the total structure function $\tilde{H}^N(s)$ in Fig. 7 d.

125

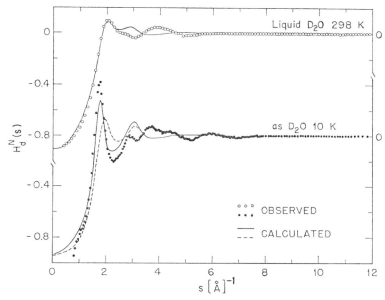

Fig. 7e. Distinct neutron structure functions, $\tilde{H}_d^N(s)$, for amorphous solid (...) and for liquid D_2O. Calculated curves are for randomly oriented water molecules with molecular center correlations derived from X-ray diffraction. (From Ref. [27])

With the molecular structure thus determined, NVR subtract the computed function $\tilde{H}_m^N(s)$ from the total structure function [Eq. (3.6)] and obtain the distinct structure function $\tilde{H}_d^N(s)$. The resulting curve is shown in Fig. 7e. The function $\tilde{H}_d^N(s)$ is related to the structure functions descriptive of atom pair interactions in $D_2O(as)$ by Eq. (3.4). With the structure functions $\tilde{h}_{OO}(s)$ known from X-ray diffraction, it is instructive to compute a neutron structure function.

$$\tilde{H}_0^N(s) = \tilde{h}_{OO}(s) \left[f_O + 2 f_D \, e^{-l_{OD}^2 s^2/2} \, \frac{\sin sR_{OD}}{R_{OD}} \right]^2 \left[\sum_\alpha |f_\alpha| \right]^{-2}. \qquad (3.9)$$

The function $\tilde{H}_0^N(s)$ describes the neutron scattering from randomly oriented D_2O molecules with the positional correlation between molecular centers specified by the correlation function $h_{OO}(R)$. We expect the orientational correlation between pairs of water molecules to be of much shorter range than the positional correlation between molecular centers, and hence the functions $\tilde{H}_0^N(s)$ and $\tilde{H}_d^N(s)$ should be nearly equal for values of $s < 2$ Å^{-1}.

The function $\tilde{H}_d^N(s)$ derived from the neutron data for $D_2O(as)$ at 10 K (points) is compared with the computed functions $\tilde{H}_0^N(s)$ in Fig. 7e. The corresponding curves for liquid D_2O are also shown. The function $\tilde{H}_0^N(s)$ computed with the function $\tilde{h}_{OO}(s)$ obtained for $H_2O(as)$ 10 K/10 K (dashed curve in Fig. 7e) does not agree with the curve $\tilde{H}_d^N(s)$ for $D_2O(as)$ at all. In particular, one cannot explain the difference in the position of the first maximum, namely 1.75 Å^{-1} in the curve

126

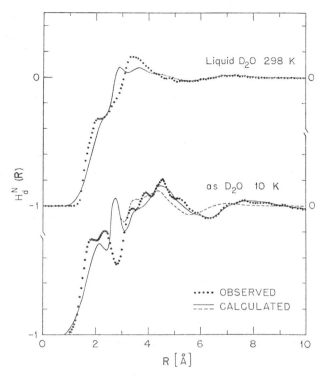

Fig. 7f. Neutron correlation functions for amorphous solid and for liquid D_2O obtained by Fourier inversion of curves shown in Fig. 7(e). (From Ref. [27])

derived from neutron diffraction and 1.95 Å^{-1} in the computed curve. This difference is not observed in the function $\tilde{H}_0^N(s)$ computed with $\tilde{h}_{OO}(s)$ taken for H_2O(as) 77 K/77 K (solid curve in Fig. 7e). We conclude that the neutron data for D_2O(as) are not consistent with results obtained from X-ray diffraction for H_2O(as) 10 K/10 K (and H_2O(as) 10 K/77 K), but are at least qualitatively consistent with the X-ray results for H_2O(as) 77 K/77 K. This is also shown in the comparison of direct space correlation functions (Fig. 7f). If we accept the conclusion that the radial distribution of oxygen atoms in D_2O(as) is consistent with that found for the high-temperature modification of H_2O(as) then it follows from Fig. 7f that the orientational correlation in D_2O(as) extends only to nearest neighbor water molecule pairs, second and higher neighbors being randomly oriented.

We conclude our discussion about orientational correlation between pairs of nearneighbor molecules in D_2O(as) with some observations about hydrogen bond distances and angles. In Fig. 7g we show the correlation functions

$$H_d^N(R) \equiv \frac{1}{2\pi^2 \varrho R} \int\limits_0^{S_{\max}} s\, \tilde{H}_d^N(s) \sin sR\, ds \qquad (3.10)$$

127

derived from neutron diffraction. Since the functions $H_d^N(R)$ represent the weighted sum of the atom pair correlation functions $h_{OO}(R)$, $h_{OD}(R)$, and $h_{DD}(R)$ with the same weight factors as in Eq. (3.4), it is instructive to subtract the contribution from oxygen atom pair interactions (dashed curves in Fig. 7g). The difference curve (solid line in Fig. 7g) represents the nearly evenly weighted sum of the functions $h_{OD}(R)$ and $h_{DD}(R)$. The function $h_{OO}(R)$ chosen for this separation was that for H_2O(as) 77 K/77 K, but our discussion concerns only the distance region $R < 3$ Å and is hence not affected by this choice. In the region of distances $R < 3$ Å the difference curves (solid lines in Fig. 7g) for liquid and amorphous solid D_2O are strikingly similar, showing a double maximum which is barely resolved in the curve for liquid D_2O, with better resolution in the curve for D_2O(as). The double maximum centered around 2 Å in the difference curve for liquid water has been interpreted as the superposition of the hydrogen bonded $O \ldots D$ distance at 1.92 Å and the nearest intermolecular $D \ldots D$ distance at 2.40 Å. The double maximum in the difference curve for D_2O(as) is barely resolved into two

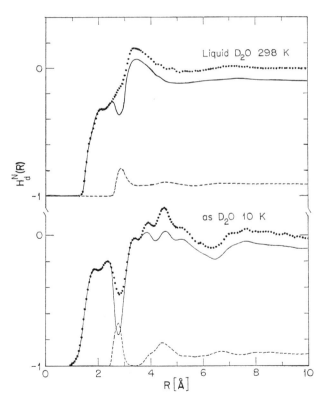

Fig. 7g. Atom pair correlation functions for amorphous solid and liquid D_2O. Dotted curves are weighted sums of contributions from OO, OD and DD interactions. Broken curves show contributions from OO interactions alone. Solid curves represent nearly evenly weighted sums of OD and DD atom pair correlation functions. (From Ref. [27])

separate peaks at 1.76 Å and at 2.30 Å. An O...D distance of 1.76 Å, when combined with an O...O distance of 2.76 Å and an intramolecular OD distance of 1.00 Å indicates linear hydrogen bonds on the average in D_2O(as) (as was found for liquid D_2O). If we assume an average OOO angle of 109.5° and an average DOD angle of 104.5°, the resulting average hydrogen-bond angle O...DO would be 2.5°, yielding an O...D distance of 1.762 Å. We estimate that a hydrogen-bond angle greater than 5° on the average is incompatible with the diffraction data on low density solid amorphous water.

IV. Vibrational Spectroscopy

The key element in all interpretations of the behavior of water is hydrogen bonding. Now, it has been known for many years that the existence of hydrogen bonding between two groups alters to some extent the vibrational motions of those groups. It is natural, then, to attempt to use studies of vibrational motion to ascertain whether or not hydrogen bonding is present, and to infer something of its relationship to local structure. Although the goal is easily stated, its achievement is very difficult.

The model fundamental to all analyses of vibrational motion requires that the atoms in the system oscillate with small amplitude about some defined set of equilibrium positions. The Hamiltonian describing this motion is customarily taken to be quadratic in the atomic displacements, hence in principle a set of normal modes can be found; in terms of these normal modes both the kinetic energy and the potential energy of the system are diagonal. The interaction of the system with electromagnetic radiation, *i.e.* excitation of specific normal modes of vibration, is then governed by selection rules which depend on features of the microscopic symmetry. It is well known that this model can be worked out in detail for small molecules and for crystalline solids. In some very favorable simple cases the effects of anharmonicity can be accounted for, provided they are not too large.

Molecules, in general, have some nontrivial symmetry which simplifies mathematical analysis of the vibrational spectrum. Even when this is not the case, the number of atoms is often sufficiently small that brute force numerical solution using a digital computer provides the information wanted. Of course, crystals have translational symmetry between unit cells, and other elements of symmetry within a unit cell. For such a periodic structure the Hamiltonian matrix has a recurrent pattern, so the problem of calculating its eigenvectors and eigenvalues can be reduced to one associated with a much smaller matrix (*i.e.* much smaller than $3N \times 3N$ where N is the number of atoms in the crystal).

Liquids do not have the symmetry characteristic of a crystal. Moreover, a molecule in a liquid is usually in an asymmetric field which, in principle, reduces or destroys the symmetry characteristic of the isolated molecule. Consequently, the program that works for molecules and crystals, does not work for a liquid.

Despite the difficulty cited, the study of the vibrational spectrum of a liquid is useful to the extent that it is possible to separate intramolecular and intermolecular modes of motion. It is now well established that the presence of disorder in a system can lead to localization of vibrational modes [28-34], and that this localization is more pronounced the higher the vibrational frequency. It is also well established that there are low frequency coherent (phonon-like) excitations in a disordered material [35,36]. These excitations are, however, heavily damped by virtue of the structural irregularities and the coupling between single molecule diffusive motion and collective motion of groups of atoms.

A. Liquid Water

In the case of water (H_2O), the free molecule frequencies are [37] $\nu_1 = 3652$ cm^{-1}, $\nu_2 = 1595$ cm^{-1} and $\nu_3 = 3756$ cm^{-1} (see Fig. 8), all of which lie very far above

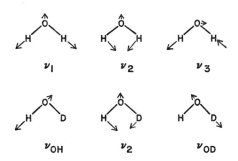

Fig. 8. The normal modes of H_2O and HOD

the range of possible translational frequencies. Hence we expect these intramolec ular motions to be strongly localized, and therefore sensitive to the local environment of a water molecule. This expectation is confirmed [38,39]; in the liquid phase there is intense absorption at $\sim$1645 cm^{-1} (ν_2) and $\sim$3500 cm^{-1} (see later for assignment). As to intermolecular modes, the rigid body librational motion in water is spread over a very wide frequency range, $\sim$300 cm^{-1} to $\sim$900 cm^{-1}, and the most prominent translational frequencies are $\sim$60 cm^{-1} and $\sim$160 cm^{-1}.

Despite the range of frequencies mentioned the observed vibrational spectrum of liquid water has broad, only poorly defined features, and some form of spectral decomposition must be adopted to enhance the retrieval of information from it. Whether one regards the spectrum as representative of superposed transitions from a continuum of environments, or from a few "chemically defined" species, coupled with selection rule breakdown, model building is enhanced by resolution of the observed band contour into independent contributions. The most commonly used technique for extracting information from the poorly resolved water spectrum is synthesis of the observed band contour from a superposition of Gaussian components [39]. The assumption that the component bands are symmetric implies that the distribution of nearest neighbor distances cannot be grossly asymmetric and that the intensity, through $\partial \alpha / \partial q$ and/or $\partial \mu / \partial q$ [40], does not vary rapidly with intermolecular displacement. It is not known if these conditions are met but, *a posteriori*, the consequences of the analysis seem reasonable, at least in terms of one model.

Several extensive reviews of the Raman [39] and infrared spectroscopy [38] of water have been published; we refer the reader to these and the original liter-ature for details. Herein we present only those data pertinent to later arguments.

Intermolecular motion: Table 1, taken from Ref. [39], displays the several inter-molecular frequencies observed using Raman, infrared, inelastic neutron and hyper-Raman spectroscopic methods. Aside from remarking that these data are con-

Table 1. Selected values of intermolecular vibrational frequencies obtained from liquid H_2O and D_2O

Raman scattering		Infrared absorption	Inelastic harmonic light scattering (Hyper-Raman)		Neutron inelastic scattering	Stimulated Raman scattering
cm^{-1}	Polarization		β_1	β_2		
H_2O						
$\sim$60	—	$\sim$70	—	—	60	—
170	p	170	—	—	175	—
425—450	dp	—	Absent	470	454	—
550	dp		535	530	590	550
		685				
720—740	dp		755	715	766	—
D_2O						
$\sim$60	—	—	—	—	$\sim$50	—
$\sim$175	—	165	—	—	$\sim$160	—
350	—	—	Absent	305	—	—
			415	400		
500	—	505			$\sim$465	—
			570	535		

sistent with the existence of local tetrahedral coordination of a water molecule in the sense defined earlier, we defer discussion of possible structural implications to a later section.

Intramolecular motion: There have been many studies of the OH stretching region of liquid water; the results of these investigations are in substantial but not complete agreement with one another. Putting aside differences for the moment, typical spectra are shown in Fig. 9. It is interesting that decomposition of the observed Raman and infrared band contours into Gaussian subcomponents gives the consistent results displayed in Table 2.

In the spectrum of pure H_2O (or D_2O), the intermolecular coupling which broadens the free molecule transitions ν_1 and ν_3 leads to severe overlap of these and (possibly) of the overtone $2\nu_2$, thereby complicating interpretation of the observed spectrum. Relatively effective decoupling may be brought about by partial isotopic substitution. In a solution HOD/H_2O or HOD/D_2O, where either

Table 2. Positions and FWHM of gaussian sub-bands in the OH stretching region of H_2O

	1	2	3	4
$\nu(cm^{-1})$	3622[1]), 3620[2])	3535[1]), 3540[2])	3455[1]), 3435[2])	3247[1]) 3240[2])
FWHM (cm^{-1})	141[1]), 140[2])	144[1]), 150[2])	176[1]), 260[2])	233[1]), 310[2])

[1]) Raman.
[2]) Infrared.

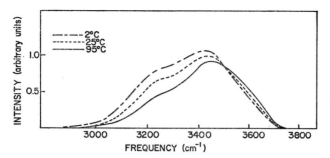

Fig. 9a. The infrared spectrum of liquid water (from Ref. [38])

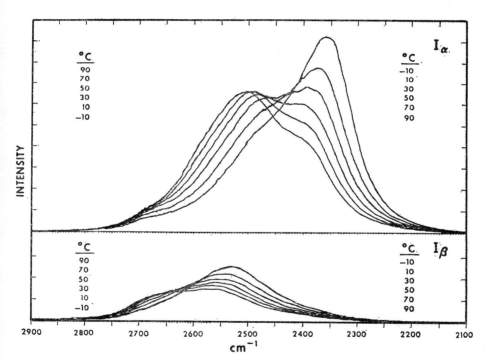

Fig. 9b. The Raman spectrum of liquid D_2O (from Ref. [42])

$$I \text{ (observed}/[y(z\,z)\,x)] = I_\alpha + I_\beta$$
$$I \text{ (observed}/[y(z\,y)\,x)] = (3/4)\,I_\beta$$

the OD or OH groups, respectively, are present as traces, the corresponding OD and OH stretching transitions are narrower than in the pure liquids D_2O and H_2O. Furthermore, ν_{OH} and ν_{OD}, derived from ν_1 and ν_3 of $H_2O(D_2O)$, no longer overlap in the fundamental region of the spectrum. A typical spectrum is shown in Fig. 10. It is clear that even in this decoupled spectrum the band contour is not symmetric. Spectral decomposition into Gaussian components, as before, leads to the parameters displayed in Table 3.

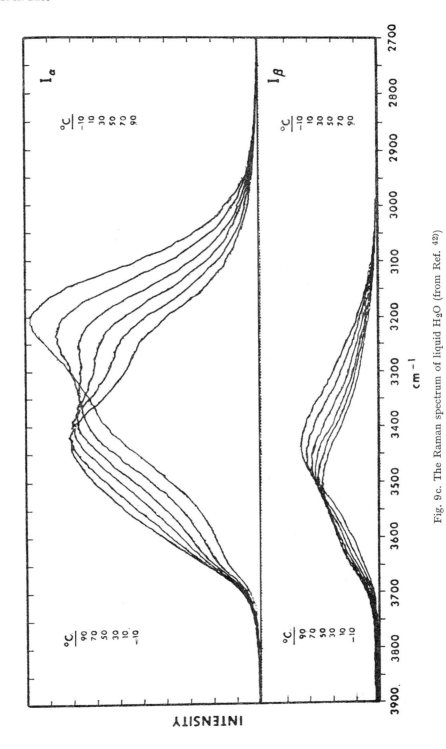

Fig. 9c. The Raman spectrum of liquid H_2O (from Ref. [42])

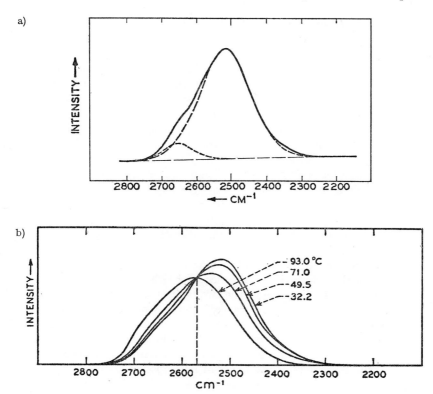

Fig. 10a and b. The Raman spectrum of liquid HOD/D₂O [from Walrafen, G.: J. Chem. Phys. *48*, 244 (1968)]

Table 3. Positions and FWHM of gaussian sub-bands in the OH (OD) stretching region of HDO/H₂O, HDO/D₂O

	1	2	1	2
ν(cm^{-1})	3435	3628	2530	2650
FWHM (cm^{-1})	260	90	145	80

Given the nature of the decomposition of an observed band contour into components, it is not surprising that there is some ambiguity with respect to the number, locations and intensities of underlying transitions [41,42]. There have been some reports of spectral decompositions into more components than we have cited, but these extra components are usually of very low intensity and spread over a large frequency range [43]. The general features of the analysis of the band contour described above are supported by studies of the overtone and combination band region of the spectrum [38], and by studies of stimulated Raman spectra [44]. We shall accept the entries in Tables 2 and 3 as a reasonable description of the observed band contours.

Note added in proof. Stimulated Raman spectra of H_2O, D_2O, H_2O-D_2O mixtures and aqueous solutions of $NaClO_4$ have been measured under conditions designed to avoid nonlinear optical effects of unwanted nature. (M. G. Sceats, S. A. Rice and J. E. Butler, J. Chem. Phys. in press). Spectra with relatively narrow peaks were obtained in all cases, as predicted by the first order theory of stimulated Raman scattering. Other effects (such as multiple peaks, broadened spectra, and shifted peaks), which have been interpreted as evidence for a mixture model of water, are reconsidered, and it is argued that they are either artifacts arising from modulation by stimulation in the hydrogen bond bending region or arise from disruption of thermal equilibrium in filaments due to the high electric fields present in the focused laser beam. The investigators conclude that information about the microscopic structure of water cannot be obtained from the stimulated Raman spectrum. The results neither contradict nor support any model of liquid water.

The temperature dependence of the vibrational spectrum of liquid water has been extensively studied [38,39]. It is found that the intensities of all transitions assigned to intermolecular modes (Table 2) decrease as the temperature increases, hence so also do the intensities of the deduced Gaussian subbands. In the OH stretching region of pure liquid H_2O the two highest frequency subbands increase in intensity and the two lowest frequency subbands decrease in intensity as the temperature is raised. The frequencies of these subbands are, however, sensibly independent of temperature. Correspondingly, in liquid HDO/H_2O the observed maximum of the OD band contour shifts to higher frequency as the temperature rises, as does the frequency of the lower of its Gaussian components, whilst the frequency of its upper Gaussian component is sensibly independent of temperature [39].

Aside from relatively small differences between the parameters of the subbands into which a given band contour has been decomposed, the major disagreement between the several sets of investigators [39,41,42,45] concerns the interpretation of the presence of an isosbestic point in superposed spectra of the same region, each spectrum referring to a different temperature. Wahlrafen's Raman data, and Senior and Verrall's infrared data [45] exhibit one, and this fact is interpreted by some to be evidence for an equilibrium between different species. Scherer, Go and Kint [42] show that the separation of the Raman spectrum into isotropic and anisotropic parts removes the isosbestic point. The existence of isosbestic points in the spectra of systems where species variation via an internal equilibrium cannot occur [46], and the generation of such points in theoretical spectra [47] without consideration of coexisting species, argue that the presence (or absence) of an isosbestic point is not necessarily significant evidence for compositional variation via chemical equilibrium.

B. Amorphous Solid Water

Many of the studies of the vibrational spectrum of crystalline ice, prepared in thin films between cooled windows, have inadvertently dealt in part with the properties of $H_2O(as)$. The most complete study, by Hardin and Harvey [48], used samples prepared under conditions designed to yield material free of crystalline ice. A less complete study of the infrared spectrum of H_2O(as) has been published by Buontempo [49]; he reported observations of only the OH stretching region in HDO/D_2O(as). Both the work of Hardin and Harvey and of Buontempo dealt, presumably, with the high temperature form of H_2O(as).

Note added in proof. It is not certain that the H_2O(as) studied by Hardin and Harvey and Buontempo is the low density form. Given the sensitivity of the form of deposit to conditions such as deposition rate and surface temperature, it is conceivable that it was the high density form. Indeed, V. Mazzacurati and M. Nardone, Chem. Phys. Lett. *32*, 99 (1975) studied the Raman spectrum of H_2O(as) deposited at 110 K. Although their sample was heavily contaminated with crystalline ice, an estimated spectral contour for H_2O(as) can be obtained. This contour, which if correct is almost certainly that of the low density form, is very different from that reported by Li and Devlin and by Venkatesh, Rice and Bates (see following text).

In general, the infrared spectrum of H_2O(as), like that of liquid H_2O, has broad, poorly differentiated peaks. Nevertheless, an internally consistent set of assignments can be made, as shown by the results of Hardin and Harvey displayed in Table 4. As will be seen in the following, the entries in Table 4 are not in good agreement with the corresponding values from Raman spectra.

Venkatesh, Rice and Bates [8] have studied the Raman spectra of H_2O(as), D_2O(as) and HDO/D_2O(as) from ∼500—4000 cm^{-1} over the temperature range ∼30 K to ∼130 K. Since their samples were deposited at ∼30 K, they presumably are of the low temperature form of H_2O(as). Figures 11 and 12 show typical spectra

Table 4. Assignments of infra-red spectral features of H_2O(as), D_2O(as) and HDO/H_2O(as), D_2O at 93 K [20] (all frequencies in cm^{-1})

Assignment[1]	Spectral character[2]	H_2O(as)	D_2O(as)	HDO/D_2O(as)	HDO/H_2O(as)
$\nu_1 + \nu_T$	Sh	3367 ± 7	2499	—	—
ν_3	VS	3253 ± 5	2436		
ν_1	Sh	3191 ± 7	2372	3304 ± 1	2437 ± 1
ν_L	W	2220 ± 5	1617	—	—
ν_2	m	1660 ± 5	1212		
ν_L	Sh	1570	—	—	
$\nu_L + \nu_T'$	SSh	846 ± 7	—	—	
ν_L	S	802 ± 5	600	792 ± 1 [3]	
$\nu_L - \nu_T'$		—	—	—	
	Sh	675 ± 7			
	mSh	535 ± 7			
ν_T		213 ± 0.5 [4]			

[1] ν_1 = OH symmetric stretching mode
ν_2 = HOH bending mode
ν_3 = OH antisymmetric stretching mode
ν_L = librational mode
ν_T = lattice mode
ν_T' = lattice mode

[2] Sh = shoulder
VS = very strong
W = weak
m = medium
SSh = strong shoulder
S = strong
msh = medium shoulder
[3] Spectral character W.

[4] The high frequency limit of this broad band is at 326 cm^{-1}, and there are shoulders at 301 and 271 cm^{-1}.

and their decomposition into Gaussian components. The experimental data are most reliable in the OH stretching region, to which we will give most attention.

VRB find a weak broad band from $\sim$500–900 cm^{-1}, which is assigned to the librational modes ν_L, possibly with some admixture of the combination bands $\nu_T + \nu_L$, where ν_T is the frequency of a hindered translational mode. They assign the next higher frequency band, stretching from $\sim$1300–1800 cm^{-1}, with peak between 1680 and 1700 cm^{-1} (H_2O(as)), to the HOH bend ν_2. In liquid H_2O and ice Ih ν_2 is 1640–1650 cm^{-1}. The VRB value is in fair agreement with the 1660 $\pm$5 cm^{-1} assigned to ν_2 by Hardin and Harvey from their infrared study of H_2O(as) [48]. Although the observed bending region band widths in both infrared and Raman studies exceed that found for the liquid [39], and the relative intensities of ν_2 and ν_L seem different for liquid and H_2O(as), uncertainties in the experimental data preclude obtaining much of value from speculations about the differences. It is to be noted, however, that there is no apparent change of frequency of ν_2 with increasing temperature.

VRB also observed a weak broad band in the region $\sim$2000–2900 cm^{-1} (H_2O(as)) and $\sim$1450–1900 cm^{-1} (D_2O(as)). The counterpart band exists in both liquid water and ice Ih and is usually assigned to the combination mode $\nu_2 + \nu_L$. This band does not exhibit a significant shift in frequency as the temperature is raised.

As shown in the figures, the spectral contour in the OH stretching region can be represented as the sum of contribution from four Gaussian components. Since

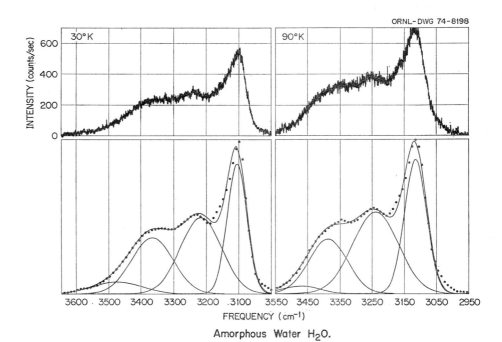

ORNL-DWG 74-8198

Amorphous Water H_2O.

Fig. 11. The Raman spectrum of H_2O(as). (From Ref. [8])

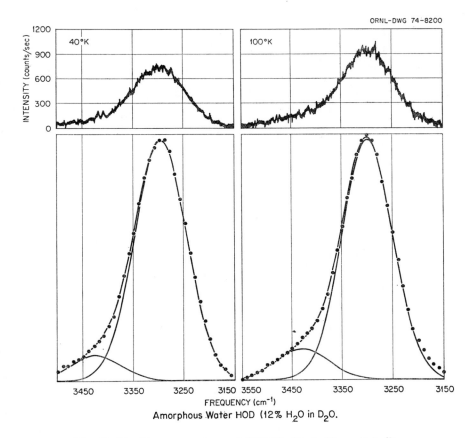

ORNL-DWG 74-8200

Amorphous Water HOD (12% H_2O in D_2O.

Fig. 12. The Raman spectrum of HDO/D_2O(as). (From Ref. [8])

the OH stretching mode band of HDO (Fig. 12) has only very slight asymmetry, and evidence for a symmetric nearest neighbor distance distribution is provided by the observed OO distribution function, in which the nearest neighbor peak is narrow and well resolved, a decomposition of the observed profile into symmetric subbands seems justified.

Tables 5 and 6 gives the temperature variations of the parameters of the component Gaussian bands in the stretching mode regions of H_2O(as), D_2O(as), and HDO/D_2O(as). As is evident from the entries, the band centers and half-widths of the component Gaussians remain nearly constant.

Although the total integrated intensities of the band envelopes apparently increase with increasing temperature, with the greatest change between 30 K and 50 K, this trend is unambiguous only for the strongest component in each of the spectra and, even in these cases, the overall change is small. The areas of the weaker bands tend to fluctuate about a mean value. The intensity changes are probably caused by changes in the sample surface.

Table 5. Temperature variation of Gaussian component parameters in the OH stretching region of amorphous solid H_2O and in the OD stretching region of amorphous solid D_2O[8]

Temp.	Band center (cm⁻¹)				Peak height[1]				Band width (cm⁻¹)[2]				Band area[1]			
(K)	1	2	3	4	1	2	3	4	1	2	3	4	1	2	3	4
H_2O(as)																
30	3108	3220	3371	3477	1.865	1.089	0.800	0.170	71	149	150	178	141	173	128	32
48	3107	3212	3357	3458	2.730	1.541	1.243	0.182	66	134	161	152	192	220	213	30
70	3112	3213	3362	3452	2.320	1.340	1.120	0.146	76	157	169	121	189	224	201	19
90	3116	3240	3388	3473	2.390	1.450	0.970	0.135	81	167	140	131	205	258	144	19
110	3116	3220	3372	3483	2.520	1.570	1.270	0.165	76	155	159	123	204	259	216	22
130	3121	3230	3362	3456	2.310	1.380	1.170	0.289	89	138	151	116	218	202	187	36
D_2O(as)																
40	2310	2358	2476	2476	1.350	0.630	0.690	0.102	45	93	125	137	65	62	91	15
60	2313	2373	2474	2524	5.800	2.600	2.620	0.731	46	89	99	144	286	248	275	112
80	2315	2374	2478	2502	5.800	2.450	2.820	0.500	52	98	111	166	320	254	333	88
100	2318	2381	2483	2530	6.450	2.830	3.000	0.580	51	112	103	172	350	338	328	106
121	2320	2384	2478	2524	7.150	2.810	3.450	0.560	53	89	106	164	402	264	388	97

1) Arbitrary units.
2) Full width at one half band maximum.

Table 6. Temperature variation of Gaussian component parameters in the OH stretching region of HDO in amorphous solid D_2O[8]

Temp. (K)	Band center (cm⁻¹) [1]		Peak height[1]		Band width (cm⁻¹)		Band area[1]	
	1	2	1	2	1	2	1	2
40	3296	3423	0.735	0.0783	120	115	94.0	9.6
60	3297	3429	0.784	0.0997	118	105	98.4	11.4
80	3306	3405	0.80	0.0593	116	110	98.8	6.9
100	3300	3430	0.93	0.1220	118	123	117.1	16.0
121	3298	3400	0.97	0.1180	110	150	114.0	18.8

[1] Arbitrary units.

Given the nature of the Raman spectrum in the OH stretching region, and the fact that the spectral decomposition used is not unique, any set of assignments for the several components must be considered tentative. The following, based on a "common sense" analysis, seems reasonable:

(i) The intense bands at ∼3112 cm⁻¹ (H_2O(as)) and 2315 cm⁻¹ [D_2O(as)] are attributed to the symmetric OH and OD stretches, respectively, of strongly hydrogen bonded species and will hereafter be denoted by ν_1^{sb} [50]. These assignments are supported by the results of Li and Devlin [51] who reported polarization effects in Raman spectra of thin films of H_2O(as) using a trapped laser beam technique. The band contour in the OH stretching mode region reported by them for $z(xx)y$ scattering is identical with the one observed by VRB for $z(x_x^z)y$ scattering. Further, for $z(xz)y$ scattering, the Li and Devlin spectrum shows a drastic reduction in the intensity of component one, suggesting it to be due to a totally symmetric mode.

(ii) The second components with maxima at ∼3220 cm⁻¹ [H_2O(as)] and 2370 cm⁻¹ [D_2O(as)] are assigned to the antisymmetric OH and OD stretches (ν_3^{sb}), respectively, of the strongly-bonded species. These assignments are in good agreement with the results of Hardin and Harvey [48] in which a strong band at 3253 cm⁻¹ in the infrared spectrum of H_2O(as) was attributed to ν_3.

(iii) The third components at ∼3370 cm⁻¹ [H_2O(as)] and ∼2475 cm⁻¹ [D_2O(as)] are assigned, based on the following considerations, to the ν_1 mode of H_2O molecules with bent hydrogen bonds (ν_1^{bb}). Noting the similarity in nearest neighbor OO separations in H_2O(as) ice Ih, ice II, and ice III, one can construct a correlation between OH and OD stretching frequency and hydrogen bond donor angle, using the notion that increasing values of ν_1 correspond to increased bending of the bond donor angle. This assumption will be discussed later (see Fig. 48). It leads to the observation that the frequency of component three corresponds to a donor angle of ∼87°. Note that the use of the described correlation implies that, in H_2O(as) deposited at low temperatures, there are bent hydrogen bonds such as those found in ices II and III. This implication is not too surprising since the low temperature deposit of H_2O(as) has about the same density as ice II and ice III. A band at ∼3367 cm⁻¹ was observed in the infrared spectrum of

141

H_2O(as) [48] and was assigned to a combination mode of ν_1 with a translational external mode. The intensity of the Raman component at 3360 cm^{-1} seems too large to be due to such a combination mode.

Li and Devlin [51] assigned the 3220 cm^{-1} band of H_2O to $2\nu_2$. Indeed, for the case of H_2O(as) the areas of bands 1 and 3 are about equal and smaller than that of band 2, and $2\nu_2$ might be expected to lie near 3250 cm^{-1}, *i.e.* under band 2. Similarly, for the case of D_2O(as) the areas of bands 1 and 2 are nearly equal and smaller than that of band 3, and in this case $2\nu_2 \approx 2450$ cm^{-1} might lie under band 3. On the other hand, the limited depolarization data [51] do not suggest that band 2 in H_2O(as) has much contribution from $2\nu_2$. The assignments given by VRB must be considered both tentative and speculative; we prefer them to assignments which invoke substantial intensity in $2\nu_2$, but cannot eliminate the possibility that some portion of the observed intensities is derived from $2\nu_2$ (presumably by virtue of interaction with ν_1) [52,53].

(iv) The fourth, weak, high frequency component ~ 3477 cm^{-1} [H_2O(as)], ~ 2476 cm^{-1} [D_2O(as)] is assigned to the symmetric stretch of a weakly hydrogen bonded species and will be denoted by ν_1^{wb}. The molecules giving rise to this band may be weakly hydrogen bonded through either or both protons. In the latter case the antisymmetric stretch would be too weak to show up on the high frequency side in the Raman spectrum. It is to be noted that even if the C_{2v} symmetry of a symmetrically bonded species, O...HOH...O, is broken by having one of the bonds weaker than the other, the stretches of the weakly and strongly bonded OH's can still be classified as symmetric, because the O...HOH...O complex has C_s point symmetry.

(v) The two components in the OH stretching region of HDO in D_2O(as) differ in intensity with the strong component centered at ~ 3300 cm^{-1} and the weak one at ~ 3415 cm^{-1} [54]. The latter is far removed from $2\nu_2$ of HDO and, therefore, cannot be identified with the overtone of the bend. Thus, the 3415 cm^{-1} component is assigned to the symmetric stretch of a weakly hydrogen-bonded species as in (d) above. The full width at half maximum (δ) of the 3300 cm^{-1} component is ~ 120 cm^{-1} as compared to $\delta \approx 70$ cm^{-1} for the ν_1^{sb} band in H_2O(as). This apparent large increase in δ for ν_1^{sb} of HDO in D_2O(as) may be caused by overlap of the three components ν_1^{sb}, ν_1^{bb} and ν_1^{wb} postulated in the interpretation of the decomposition of the H_2O(as) Raman spectrum. As a test of this supposition VRB made a least squares fit to the observed band contour using three Gaussian functions (recall that for HDO ν_3 is displaced). They find a best fit with component frequencies considerably shifted from those of H_2O(as). The width of the best fit composite band is ~ 118 cm^{-1}, to be compared with the observed value ~ 134 cm^{-1}. The peak positions of the two larger components are so close (3296, 3303 cm^{-1}) that the result is really a two component fit, where one component has a symmetric but not simple Gaussian shape. VRB are unable to account for the band shifts required if this interpretation is driven to its logical conclusion. Perhaps the local structure of HDO/D_2O(as) is not the same as that of H_2O(as). Certainly, the rate of relaxation of a "hot" HDO molecule after landing on a surface of D_2O(as) will be different from that of a "hot" D_2O molecule [55] and this could lead to generation of different environments for the two cases. The increase in component band widths would, then, represent a measure of the increased heterogeneity of HDO environments relative to the distribution characteristic of pure H_2O(as).

Note added in proof. Kamb, Hamilton, LaPloca and Prakash [J. Chem. Phys. *55*, 1934 (1971)] have argued, from the comparison of the infrared absorption spectrum of HDO/H_2O ice II and crystal structure data, that there is no correlation between ν_{OD} and hydrogen bond donor angle. Instead, for this case they establish a correspondence between ν_{OD} and OO separation. This interpretation differs from that offered by VRB for high density H_2O(as).

Now, in high density H_2O(as) the rms width of the first neighbor peak of $g_{OO}(R)$ is 0.135 Å.[27] If the frequency of OH stretching depended only on OO separation we would expect an essentially Gaussian spectral contour, since the nearest neighbor OO separation is well described by a Gaussian distribution of distances. Using HDO/H_2O ice II data, the rms width of such a spectral distribution of ν_{OD} in HDO/H_2O(as) would be of the order of 70 cm^{-1}. For the corresponding ν_{OH} in HDO/D_2O(as) we would expect an rms width of the order of 150—200 cm^{-1}, using the ratio of widths δ_{OD}/δ_{OH} for HDO/H_2O, D_2O ice II.[1] All in all, it is conceivable that the shape of the dominant Gaussian component of HDO/D_2O(as) reflects just the natural OO distribution. However, even given that ambiguities associated with decomposing a curve into Gaussian subbands can lead to spurious components, the fact that the nearest neighbor OO distribution is Gaussian for H_2O(as) (and only this distribution will influence the OH stretching frequency) while the Raman spectral contour is asymmetric, suggests that it is unlikely that the VRB spectral decomposition has such spurious components. In any event, there seems no easy way to accommodate a Gaussian distribution of near neighbor OO separations and an asymmetric Raman spectral contour with a linear, or nearly linear, relationship between ν_{OH} and OO separation.

The case of pure H_2O(as) is harder to discuss, since the strong coupling of vibrations broadens the spectrum considerably. The three largest Gaussian components inferred to be part of the spectral contour of H_2O(as) have about equal area. It is hard to see how a single Gaussian distribution of OO separations, together with a nearly linear OH stretching —OO separation relation, could lead to an overall contour with this property. We conclude, then, that these components are likely "real". Furthermore, a linear correlation between OH stretching frequency and OO separation leads one to expect equal numbers of OO pairs with Gaussian distributions centered around three OO separations. Using the KHLP data for, *e.g.* ice II, these separations span the range 2.768 Å to 2.844 Å, which would lead to a first peak in $g_{OO}(R)$ much broader than what is observed.

We conclude that it is likely that the OH stretching frequency depends on both OO separation and hydrogen bond donor angle, and that neither the VRB or KHLP correlations is complete. For the present we accept the VRB association of spectral components with local environments as a conjecture which must be tested against other experimental data.

If it is accepted that Fermi resonance between 2 ν_2 and ν_1 is unimportant, and that 2 ν_2 does not appear in the spectrum, the assignments of ν_1^{bb} and ν_3^{sb} to components can be reversed at the expense of increasing the $\nu_1-\nu_3$ splitting relative to the value in the isolated molecule ($\sim$150–160 cm^{-1} versus 104 cm^{-1}).

Perhaps, in view of the intermolecular hydrogen bonding, this is not unreasonable. The advantage of this alternative assignment is that it is somewhat more consistent with the crude depolarization data of Li and Devlin, *i.e.* the polarization of ν_1^{bb} is now greater than that of ν_3^{sb}. Furthermore, if the amorphous solid is a model of the liquid and the ordering in frequency of ν_1^{bb} and ν_3^{sb} is the same in the liquid, this assignment fits better the depolarization data for the liquid.

Even if the infrared and VRB Raman studies refer, respectively, to high and low temperature forms of H_2O(as), the nearest neighbor distributions of these are so nearly the same that there is no ready explanation for the discrepancies in the assigned frequencies from infrared and Raman spectroscopic studies. It could be said that the infrared derived values are uncertain by virtue of the difficulty of separating shoulders from broad absorption contours. On the other hand, the decomposition of the Raman spectrum into Gaussian sub bands is an *ad hoc* procedure, possibly leading to systematic error. Similar discrepancies in the data sets for the several crystalline ices have been discussed by Taylor and Whalley [53], who argue for the importance of intermolecular coupling and different extents of breakdown of the free molecule selection rules for infrared and raman activity. Fortunately, the main features of the two kinds of spectra, and the assignments made, are in agreement, so that analyses of changes in a spectrum will be reliable if based on consistent use of the same kind of experimental data.

Note added in proof: V. Mazzacurati and M. Nardone [Chem. Phys. Lett. 32, 99 (1975)] have reported a brief study of Raman scattering from H_2O(as) deposited at 110 K. As can be seen from Fig. A the sample is heavily contaminated with crystalline ice and the spectral contour for H_2O(as) must be estimated. Assuming their estimation procedure is accurate, and that the resulting spectral contour for

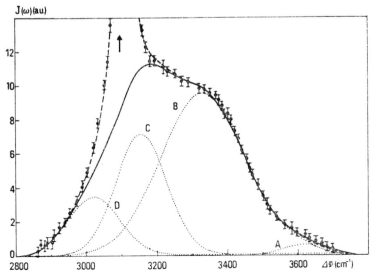

Fig. A. Background subtraction and Gaussian decomposition of the NM Raman spectrum of H_2O(as) deposited at 110 K

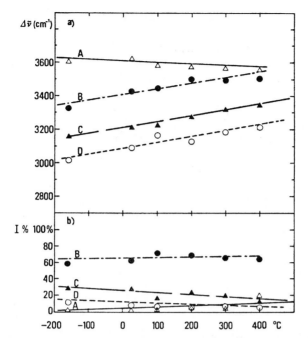

Fig. B. Frequency (a), and percentage of integrated intensity (b) versus temperature for the subcomponents of Fig. A. The data above 0 °C is obtained from Lindner's thesis [Karlsruhe (1970)].

H_2O(as) can be decomposed into Gaussian subbands, it is found that the subband frequencies and intensites fall on the extrapolations of the respective temperature dependences of the Gaussian subbands fitted to the liquid Raman spectral contour (see Fig. B). This sample of H_2O(as) is almost certainly of the low density form, so the results are consistent with the assertion that low density H_2O(as) is closely related to supercooled water, a suggestion which is discussed in Sections VI and VII.

This result is not as satisfactory as it seems. The reader will notice that the behavior of the Gaussian components appears different from that described on p. 130. MN use the data of H. A. Lindner [Ph. D. dissertation, University of Karlsruhe (1970)], and his spectral decomposition must differ from that of Walrafen judging from the frequencies (see Table 2) as well as the temperature dependences of the components (see also the comments by Walrafen [39]). Until this difference is resolved, and better spectra are available, the findings of MN must be viewed with reserve.

V. Some Features of Models of Liquid Water

A number of models of liquid water, of varying sophistication and purpose, have been proposed since the turn of the century. It has become conventional to classify these under the headings "continuum" and "mixture", a terminology designed to emphasize how the intermolecular hydrogen bonding is described. Since the various models have been reviewed from several points of view recently [1,3,57,58], they will not be described in detail here. We will, however, abstract those features of the various models that are common to all, or are of particular interest for our later considerations.

One of the most important developments in the theory of water is the invention of a realistic effective intermolecular pair potential. Although others had made similar suggestions earlier [59], it is Stillinger and co-workers [60] who have shown, with the aid of molecular dynamics calculations [3], how satisfactory such a potential can be. The availability of a water-water potential drastically reduces the scope for parameterization in model based theories, and permits investigation of such concepts as "broken" hydrogen bond. We shall frequently have occasion to call upon its properties in the following discussion.

A. Lattice and Cell Models

Given the character of the water-water interaction, particularly its strength, directionality and saturability, it is tempting to formulate a lattice model, or a cell model, of the liquid. In such models, local structure is the most important of the factors determining equilibrium properties. This structure appears when the molecular motion is defined relative to the vertices of a virtual lattice that spans the volume occupied by the liquid. In general, the translational motion of a molecule is either suppressed completely (static lattice model), or confined to the interior of a small region defined by repulsive interactions with surrounding molecules (cell model). Clearly, the nature of these models is such that they describe best those properties which are structure determined, and describe poorly those properties which, in some sense, depend on the breakdown of positional and orientational correlations between molecules.

Lattice and cell models are generally used to calculate the thermodynamic properties of a liquid. Since these, with the exception of the entropy and its derivatives, are frequently insensitive to details of local structure, the models can only be relied upon to give useful information about qualitative changes, e.g. interpreting the maximum in the density-temperature profile of liquid water. It is also possible, by examination of the internal consistency between the lattice topology and the equilibrium hydrogen bonding pattern predicted for that lattice model, to infer something about the overall importance of hydrogen bond connectivity in the liquid phase.

Recently, similar but not identical, lattice models of water have been proposed by Fleming and Gibbs [61] and by Bell [62]. In both models molecules are restricted to occupation of the sites of a body centered cubic array. The fundamental tetrahedrality of the water-water interaction is accounted for in that four noncontiguous nearest neighbor points of the total of eight nearest neighbor points of a

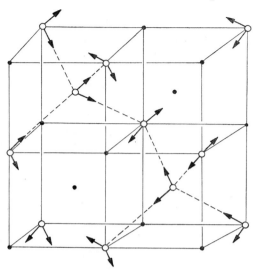

Fig. 13. Tetrahedral bonding of water molecules as related to the ice Ic structure and the bcc lattice

given lattice point are at the vertices of a tetrahedron (see Fig. 13). Indeed, ice Ic has a crystal structure which can be thought of as a body centered cubic lattice with only every other lattice site occupied. It should be noticed that two interpenetrating cubic lattices coexist on any given body centered cubic lattice [11]. If water molecules occupy the vertices of these interpenetrating lattices, two completely hydrogen bonded nets are generated. Each molecule in one net is bonded to four others in the same net, and to no molecule in the other net. The structure of ice VII is of this form [11].

In the FG and B lattice models the possible orientations of pairs of water molecules are restricted such that the two OH groups of one molecule point toward a pair of nearest neighbor cube vertices which are placed at opposite ends of a cube face diagonal. Since there are 12 nearest neighbor pairs of this type, a water molecule with distinguishable hydrogens has 24 possible orientations. In the Fleming-Gibbs version of the model only pair interactions between nearest neighbor molecules are allowed. The pair interaction itself is parameterized to have only three possible values: $-E_b$ if the orientations of the two molecules generate a hydrogen bond, $-\varepsilon_1$ if rotation of one molecule will generate a hydrogen bond, and $-\varepsilon_2$ if rotation of both molecules is required to generate a hydrogen bond. Of the $(24)^2 \equiv 576$ possible orientations of a pair of molecules that are nearest neighbors, 72 have energy $-E_b$, 360 have energy $-\varepsilon_1$, and 144 have energy $-\varepsilon_2$. The three energy parameters satisfy the condition

$$- \varepsilon_2 > - \varepsilon_1 > - E_b .$$

In the Bell version there are two molecule and three molecule interaction energies. The parameters of the interaction are: $-E_b$ if the orientations of the two molecules generate a hydrogen bond, $-\varepsilon_1$ if there is not a hydrogen bond, and an interaction $u/3$ between three simultaneously occupied sites in a $45° - 45° - 90°$ triangle the legs of which connect nearest neighbors and the hypotenuse of which connects second neighbors. There are no such triangles wholly on one of the interpenetrating sublattices of the body centered cubic lattice; all such triangles have vertices on both of the interpenetrating sublattices. The energy parameters satisfy

$$+ u > - \varepsilon_1 > - E_b .$$

In both versions of the lattice model the interaction involving non nearest neighbors, namely ε_2 or u, is introduced to destabilize occupancy of both interpenetrating sublattices relative to occupation of only one.

Calculation of the partition function, even for the simple models described, cannot be carried out exactly. For details of the approximations used the reader is referred to the original papers [61,62]. From our point of view the most interesting results are:

(i) The density-temperature profile is predicted to have a maximum.

(ii) The average number of nearest neighbors to a given molecule is relatively insensitive to temperature over the liquid range.

(iii) (Bell model) The existence of a density maximum, and a compressibility minimum, is not very sensitive to the values of the ratios

$$(\varepsilon_1/E_b - \varepsilon_1) \text{ and } u/(E_b - \varepsilon_1);$$

the extrema referred to tend to weaken as the pressure increases, just as observed.

Two cell models of water have been reported. Weissman and Blum [63] considered the motion of a water molecule in a cell generated by an expanded but perfect ice lattice. Weres and Rice [64] developed a much more detailed model, based on a more sophisticated description of the cell and a good, nonparametric water-water interaction namely the Ben-Naim-Stillinger potential [60]. The major features of the WR model are the following:

(a) Just as in the static lattice models, it is necessary to account for the multiplicity of possible arrangements of a water molecule and those that surround it. WR also choose (as did FG and B) the body centered cubic lattice as reference, but they consider a large basic lattice section which is a regular octahedron containing a central site and its fourteen nearest and next nearest neighbor sites. There is a different one molecule cell partition function corresponding to each state of occupation of the basic lattice section.

(b) The influence on the central molecule of interactions with molecules outside the basic lattice section is accounted for by a continuum approximation.

(c) Following generation of the states of occupation of the basic lattice section, and evaluation of the entropy associated with the distribution of configurations, the cell partition function corresponding to each configuration is evaluated using the BNS potential or a modification thereof.

(d) The free energy of the system is minimized with respect to the weights of the contributing basic lattice section configurations.

Clearly, the Weres-Rice model has a hybrid character. The variety of possible environments of a water molecule is approximated by the discrete set of configurations corresponding to states of occupation of the basic lattice section, and these states of occupation are counted as if each OH group can only be hydrogen bonded or not. Of course, there is a subsidiary condition, namely the "ice rule", guaranteeing that only proper molecules and bonding situations occur [65]. Having generated the possible states of occupancy of the basic lattice section, and thereby the set of possible cell environments, the notion of broken or unbroken hydrogen bonds is discarded, and the cell partition function is evaluated with a well defined, nonparametric, interaction potential.

The actual numerical evaluation of the cell partition functions, and the final minimization of the free energy are extremely involved, and several approximations were introduced by WR to simplify the calculations. For details we refer the reader to the original paper [64] and the elegant review by Stillinger [3]. The principle results are:

(i) Over the ordinary liquid range the computed thermodynamic functions have the proper trends with changing temperature, and about the right magnitude (see Figs. 14–17 and Table 7). The specific heat is the least well described thermodynamic function, which is not surprising in view of the fact that it is a temperature derivative of the entropy.

Table 7. An analysis of the values of Enthalpy and Entropy from the Weres-Rice cell model

	$t = 0\,°C$	$t = 100\,°C$
Enthalpy, kcal/mol		
Lattice	−8.296	−8.088
Translational[1]	1.729	2.290
Librational[1]	2.517	2.858
Non-hydrogen-bonded neighbor[2]	−1.325	−1.271
Long range	−1.190	−1.166
Intramolecular zero point[3]	−0.900	−0.694
Total	−7.465	−6.071
Experimental total	−8.594	−6.791
Entropy, cal/° mol		
Configurational	4.480	4.480
Orientational	1.700	1.690
Translational	7.120	9.280
Librational	2.000	3.440
Non-hydrogen-bonded neighbor[2]	−1.170	−0.950
Vibrational	0.260	0.260
Total	14.390	18.200
Experimental total	15.170	20.790

[1] Includes zero point energy.
[2] Refers to perturbation terms.
[3] Relative to vapor.

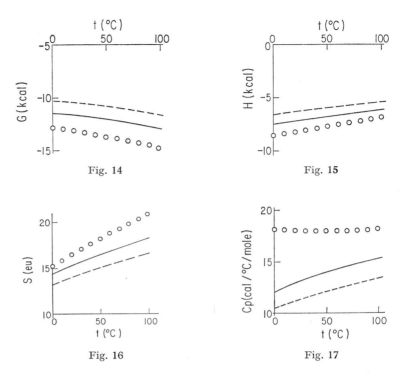

Fig. 14. The Gibbs free energy of liquid water calculated from the Weres-Rice Model (from Ref. [64]). Experiment **o**. BNS potential –––. Modified BNS potential. Corrected librational frequencies) ——

Fig. 15. The enthalpy of liquid water calculated from the Weres-Rice model (from Ref. [64])

Fig. 16. The entropy of liquid water calculated from the Weres-Rice model (from Ref. [64])

Fig. 17. The specific heat of liquid water calculated from the Weres-Rice model (from Ref. [64])

(ii) Returning for the moment to the "broken-unbroken" bond description of the basic cell occupancy states, it is found that the average number of hydrogen bonds per molecule is 1.35, sensibly independent of temperature. Indeed the contributions of occupancy states with doubly, triply and quadruply bonded water molecules are each sensibly independent of temperature.

(iii) The average number of nearest neighbors, namely 4.7, is sensibly independent of temperature and is in modestly good agreement with the value deduced from X-ray data, namely 4.4 [16].

(iv) The calculated hindered translation frequency spectrum has peaks at 70 cm^{-1} and 190 cm^{-1}, which agree well with the observed values 60 cm^{-1} and 170 cm^{-1} (see Fig. 18). Similarly, the calculated intramolecular frequency distribution [66] is at least not in disagreement with the observed spectrum (see Fig. 19). Of particular interest, in this latter case, is the "splitting" of the OH stretching mode region of the theoretical density of eigenfrequencies.

Taken together, the static lattice and cell model theories strongly suggest that:

(i) The classical "anomalies" in the density-temperature and compressibility-temperature profiles of water are primarily reflective of local tetrahedral ordering and the destabilization of more closely packed structures relative to tetrahedral ordering.

(ii) The local structure of the liquid is relatively insensitive to temperature over the ordinary liquid range.

(iii) An appropriate model of liquid water must, in some sense, involve a randomized, imperfect, but space filling network of hydrogen bonded molecules.

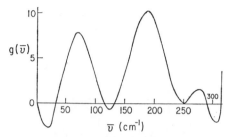

Fig. 18. The hindered translational region of the vibrational spectrum as predicted by the Weres Rice model (from Ref. [64])

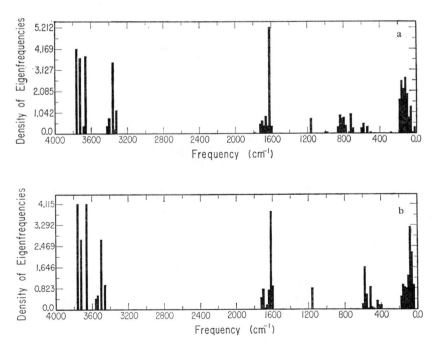

Fig. 19. The vibrational eigenvalue spectrum of the Weres-Rice model (from Ref. [66]). (a) 0 °C. (b) 100 °C

B. A Phenomenological Synthesis

The preceding general conclusions can be introduced into a phenomenological formulation which combines elements of the structural and energetic parts of the theoretical analysis. The argument, as proposed by Angell [67], goes as follows. Suppose the "ground state" configuration of water, corresponding to a hypothetical liquid state at 0 K, consists of a fully connected random tetrahedral network such as that characteristic of Ge(as) [22]. In this network each water molecule is hydrogen bonded to four others, hence is always in strong interaction with its surroundings. Now map the random tetrahedral network into a "bond lattice", by representing each water-water interaction with a point. Since every water molecule interacts with four others, there are twice as many points in the bond lattice as there are molecules in the sample (see Fig. 20). Analagous to the enumeration of occupancy states in the lattice and cell models, the topological state of the liquid can be defined by associating a value with each of the bond lattice points.

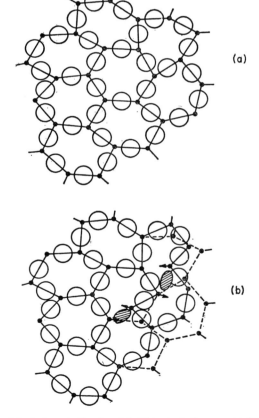

Fig. 20. The relationship between the bond lattice and the parent lattice; (a) unexcited, (b) with two excitations and local collapse (from Ref. [67])

In the simplest version of the transformation, accepting for this purpose the notion of broken and unbroken hydrogen bonds, each bond lattice point is either "occupied" or "unoccupied", respectively. "Turning off" the water-water hydrogen bond interaction is thereby mapped into an elementary configurational excitation of the bond lattice. Angell visualizes such an excitation in real space as the response of a group of molecules with strained bonding arrangement to the increasing amplitudes of motion of its members. The instability is (imagined to be) resolved by a cooperative motion in which a nearest neighbor pair of oxygens "snaps" out of line along the OH...O axis whilst the other neighbors undergo small displacements which reduce the average local strain to a minimum. This visualization is useful, but is not necessary to the definition of the model.

In the simplest version of the transformation, the state of occupancy of a given bond lattice point is independent of the states of occupancy of other bond lattice points; this corresponds to neglecting interaction between hydrogen bonds. The calculation of the distribution of possible occupation states of the bond lattice replaces the enumeration of occupancy states of the basic lattice section in the cell model, but in the simplest model the bond lattice occupancy distribution only accounts for a subset of possible basic lattice section occupancies.

If the notion of broken and unbroken hydrogen bonds is retained for the purpose of calculating the thermodynamic weights of different bond lattice occupancy states, and an energy ε_0 is assigned to the elementary configurational excitation, what would be the evaluation of the cell partition function is reduced to the calculation of the simple Boltzmann factor $\exp(-\xi\varepsilon_0/k_B T)$, $\xi=0$ or 1. The full partition function is generated by combining properly all such Boltzmann factors The results of interest are

$$\frac{n_0}{2N} = (1+e^{-\varepsilon_0/k_B T})^{-1} \tag{5.1}$$

and

$$\frac{C_v}{2Nk_B} = \left(\frac{\varepsilon_0}{k_B T}\right)^2 \frac{e^{\varepsilon_0/k_B T}}{(1+e^{\varepsilon_0/k_B T})^2} \tag{5.2}$$

$$= \left(\frac{\varepsilon_0}{k_B T}\right)^2 \frac{n_0}{2N}\left(1 - \frac{n_0}{2N}\right)$$

where n_0 is the number of unoccupied bond lattice sites (broken hydrogen bonds) and $2N$ the total number of bond lattice points (for N water molecules). The phenomenological argument can be extended to account for subsidiary one particle effects, i.e. those that can still be described without explicit particle-particle interaction, by replacing ε_0 by a free energy of excitation $f_0 \equiv \varepsilon_0 - Ts_0$. The principal difference that results from this change is the replacement of ε_0 by f_0 in the expression for C_v, whereupon C_v can be larger than before by virtue of the nonzero entropy of excitation s_0. Fits of this phenomenological model to experimental data are described by Angell [67] and will not be discussed here.

The random network bond lattice transformation very clearly displays some aspects of the relationship between the cell model and broken bond (multistate) models. We have already remarked on the analogies between states of occupancy of the basic lattice section and the distribution of occupancy in the bond lattice, and the value of the cell partition function and the simple Boltzmann weight $\exp\left(-\xi\,\varepsilon_0/k_B\,T\right)$. Clearly, accounting for interaction between bond lattice points, thereby changing the distribution of occupancy, is analogous to accounting for a larger range of the possible basic lattice section occupancies with weights, and allowing for several levels of elementary configuration excitation is analogous to improving the evaluation of the cell partition function. Just as the value of the cell partition function depends on the occupancy state, when interactions between bond lattice points are considered the weight associated with a state of occupancy must be determined self-consistently with respect to the occupancy of surrounding points.

C. The Cluster Model

The theoretical description based on the lattice or cell models of the liquid uses the language "contributing states of occupancy". Nevertheless, these states ot occupancy are not taken to be real, and the models are, fundamentally, of the continuum type. The contribution to the free energy function of different states of occupancy of the basic lattice section is analogous to the contribution to the energy of a quantum mechanical system of terms in a configuration interaction series.

Mixture models differ from continuum models by virtue of the assertion that the separate contributions to any property from the several species are, in principle, measurable. When this assertion is abandoned, the differences between models often boil down to semantic niceties, and conflicting estimates of the adequacy of the starting points.

One well worked out example of a mixture model is the cluster theory proposed by Sheraga and co-workers [68,69]. It is assumed that liquid water consists of a mixture of clusters of hydrogen bonded molecules (up to nine molecules in the latest version [69]). The clusters are constructed so as to have the maximum possible hydrogen bonding, and interactions between clusters in the "mixture" are neglected, except for hard sphere excluded volume and mean field corrections to their translational partition functions. The model includes, in the energies of the clusters, deviation from additivity of hydrogen bond interactions. Intermolecular vibrations of the molecules within any cluster are derived from the assumed geometry and the BNS potential.

Just as in our abbreviated descriptions of the lattice and cell models, we shall not be concerned with details of the approximations required to evaluate the partition function for the cluster model, nor with ways in which the model might be improved. It is sufficient to remark that with the use of two adjustable parameters (related to the frequency of librational motion of a cluster and to the shifts of the free cluster vibrational frequencies induced by the environment) Scheraga and co-workers can fit the thermodynamic functions of the liquid rather well (see Figs. 21–24). Note that the free energy is fit best, and the heat capacity worst (recall the similar difficulty in the WR results). Of more interest to us, the cluster model predicts there are very few monomeric molecules at any temperature in the normal liquid range, that the mole fraction of hydrogen bonds decreases only slowly with temperature, from 0.47 at 273 K to 0.43 at 373 K, and that the low

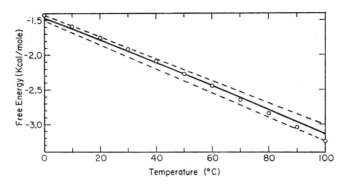

Fig. 21. The Gibbs free energy fit as given by the cluster model (from Ref. [68])

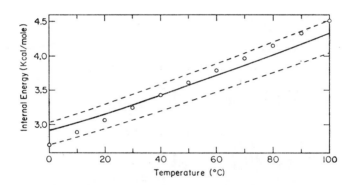

Fig. 22. The enthalpy fit as given by the cluster model (from Ref. [68])

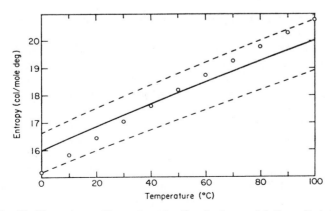

Fig. 23. The entropy fit as given by the cluster model (from Ref. [68])

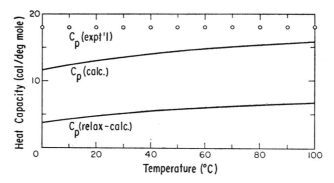

Fig. 24. The specific heat fit as given by the cluster model (from Ref. [68])

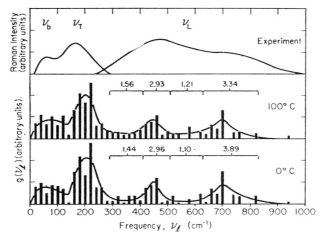

Fig. 25. The vibrational eigenvalue spectrum from the cluster model (from Ref. [68])

frequency intermolecular vibrational spectrum has peaks at ~50 cm^{-1} and ~ 210 cm^{-1} (see Fig. 25).

Note added in proof. It should be emphasized that the parametric fit used by Scheraga and co-workers is not achieved by minimizing the theoretical free energy function with respect to said parameters. Rather, the difference between the theoretical free energy function and experimental data is minimized with respect to these parameters. The method is, therefore, related to curve fitting.

Leaving aside questions of motivation, point of view, and consistency of argument, a rudimentary analogy can be made between cluster partition functions and the cell partition functions corresponding to different states of occupancy of the basic lattice section of the cell model [70]. It should not be too surprising then, that the qualitative features of the hindered translation spectrum predicted by these models are the same. They agree, also in the prediction that there is not a large structural change over the normal liquid range (though the cluster model predicts larger changes than does the cell model). It is interesting that these predictions agree despite the differences in states of occupancy (or cluster geometries) sampled in the two approaches.

D. Molecular Dynamics and Monte Carlo Computer Simulations

The most valuable of all the models of water, by far, is the computer simulated liquid with well defined water-water interaction. To date, molecular dynamics simulations for two pair potentials [3], and Monte Carlo simulations for three pair potentials [71,72], have been published. The details of the methods of simulation can be found in the literature, to which the reader is referred.

Consider, first, the sensitivity of the liquid structure to differences in two water-water potentials. Watts [72] has determined the pair functions $g_{OO}(R) \equiv h_{OO}(R) + 1, g_{OH}(R)$ and $g_{HH}(R)$ by Monte-Carlo simulation for both the Ben-Naim-Stillinger [60] and Rowlinson [59] potentials. These potentials are similar, each having a spherically symmetric core component (represented by a Lennard-Jones 6–12 form) and an angle dependent long ranged component (generated by a tetrad of charges). They are different in that the dipole moment of the BNS model (2.10 D) is larger than that of the Rowlinson model (1.84 D), and the short range divergences arising from overlap of the charge tetrads of two molecules are avoided differently (via a smooth switching function in the BNS model and via a hard sphere cutoff in the R model). Figures 26–28 show the OO, OH and HH distribution functions found. Clearly, the overall features of the distribution functions for the two model potentials are the same, but there are significant differences of detail, particularly in the OO and HH functions. For example, the predicted nearest neighbor OO peaks are at the same distance, but the amplitudes do not agree (the BNS function peak is much larger than the R function peak), and beyond the first peak the BNS function has larger amplitude oscillations (more structure) than does the R function. In the OH and HH distribution functions there are differences in peak locations and peak shapes. In general, the BNS

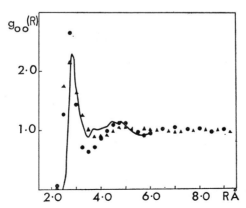

Fig. 26. The OO pair correlation function from the Monte Carlo simulation of liquid water ● BNS potential, ▲ Rowlinson potential, —— (from Ref. [72])

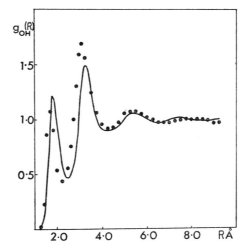

Fig. 27. The OH pair correlation function from Monte Carlo simulation of liquid water (from Ref. [72]). ● Rowlinson potential —— BNS potential

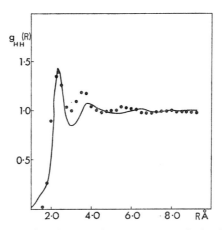

Fig. 28. The HH pair correlation function from the Monte Carlo simulation of liquid water (from Ref. [72]). ● Rowlinson potential —— BNS potential

potential leads to a structure with enhanced local tetrahedral character relative to that determined by the R potential.

Table 8 displays the computed internal energy and specific heat at constant volume. The fact that these agree well with each other, and with experiment, despite the differences in local structure of liquids based on the two potentials emphasizes the insensitivity of thermodynamic properties (except pressure) to structural details.

We conclude that care must be exercised lest characteristics of the computer simulated liquid which arise from specific details of the water-water potential used, and not from general features of that potential, be inferred to be character-

Table 8. Comparison of internal energies and heat capacities for BNS and R potentials from Monte-Carlo simulation [72]

Potential	No. of [1] molecules	U [2]	C_v [3]
R	64	−8.20[4]	21.0[4]
BNS	64	−6.80[4]	20.5[4]
Expt	—	−8.12	18.0

[1] Simulations with 216 and 500 molecules are also reported. The results do not differ significantly from those displayed.
[2] kcal/mole.
[3] cal/° mole.
[4] Including a dielectric continuum estimate of the long ranged dipole-dipole interaction, using a dielectric constant of 78.5 for the continuum outside the sample space of Monte-Carlo simulation.

istic of real water. Certainly, the general features of the model liquids must also be features of real water, but whether any of the more detailed behavior can be similarly used to analyze real water remains to be established. In particular, given the high frequency librational motion of the molecules, there must be some difference between the predictions of quantum mechanics, and those of classical mechanics on which the simulations thus far published have been based.

Having stated the principle caveat with respect to interpretation of the results of computer simulation of liquid water, we now turn to an examination of the set of results for one potential, namely the improved ST-2 potential [73]. Figure 29 displays the OO distribution function using that potential ($T = 283$ K, $\varrho_m = 1$ gm cm^{-3}), along with a comparison with the function deduced from X-ray scattering. The nearest neighbor peak locations are in excellent agreement, but the simulated liquid peak is higher and narrower. The breadth and locations of subsequent maxima in the calculated $g_{OO}(R)$ are also in agreement with the experimental data. Calculations of $g_{OO}(R)$ for different temperatures reproduce the trends observed in the X-ray data (see Fig. 2). The simulated liquid also possesses extrema in the density-temperature and compressibility-temperature profiles, just as does real water.

For completeness, we show in Fig. 30 the functions $g_{OH}(R)$ and $g_{HH}(R)$ for this simulation. In the next section we will discuss the little that can be inferred about these from diffraction data and plausibility arguments.

Of greater importance to us is the interpretation of the role of hydrogen bonding in determining the properties of simulated water. We have consistently employed the view that a hydrogen bond energy varies with length and angle, but is not "on or off". The starting point of the computer simulations, namely the ST2 potential, has this characteristic. Is it necessary to change point of view, because the effective behavior in the many body system has "on-off" attributes not obvious

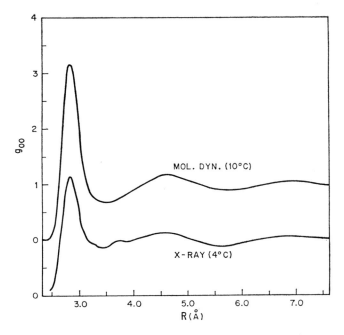

Fig. 29. The OO pair correlation function for the ST_2 potential from molecular dynamics simulation of liquid water (from Ref. [3])

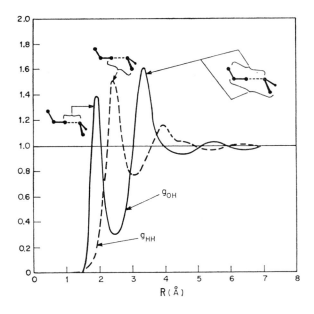

Fig. 30. The OH and HH pair correlation functions for the ST_2 potential from molecular dynamics simulation of liquid water (from Ref. [3])

in the bare potential? Stillinger and Rahman have investigated this matter by examining the classification of molecules into groups with zero, one, ..., four hydrogen bonds for a range of energies defining "on/off" for a hydrogen bond [3]. For the ST2 potential a value of V_{HB} between -1.7 kcal/mole and -4.5 kcal/mole, where a molecule is hydrogen bonded if its pair interaction is more negative than V_{HB}, and not if the pair interaction is more positive than V_{HB}, leads to the conventional hydrogen bond pattern of ice Ih. They find that for all temperatures and densities studied the distributions of hydrogen bonded molecules have a single maximum. At least for the case of simulated water it can be asserted that the interstitial model fails, since it requires a bimodal hydrogen bond distribution.

In an early paper Stillinger and Rahman computed

$$\mathscr{P}(V) = \frac{\varrho}{8\,\pi^2} \int d\boldsymbol{x}_2 \; \delta[V - V_{\mathrm{eff}}(\boldsymbol{x}_1, \boldsymbol{x}_2)] \; g(\boldsymbol{x}_1, \boldsymbol{x}_2) \,; \tag{5.3}$$

$$\boldsymbol{x} \equiv (\boldsymbol{R}, \varOmega)$$

which is the density in energy of pairs of molecules which have interaction energy V. In Eq. (5.3), $g(\boldsymbol{x}_1, \boldsymbol{x}_2)$ is the full, angle and distance dependent, pair correlation function. Aside from a trivial, but annoying, divergence as $V \to 0$ because of the contributions of well separated pairs which have a dipole-dipole interaction, $\mathscr{P}(V)$ is expected to show a dependence on V that is related to structure. For example, for ice $\mathscr{P}(V)$ will have only discrete values corresponding to pairs of molecules at the various separations allowed in the crystal lattice. Stillinger and Rahman find that $\mathscr{P}(V)$ is bimodal in simulated water. It is hard to interpret this result uniquely, since it refers to molecules with all possible relative orientations. Thus, in principle, two molecules with a "bent hydrogen bond" at separation R can have the same energy as two molecules with a perfect linear hydrogen bond but further apart. The two peaks in $\mathscr{P}(V)$ cannot then be simplistically labelled as referring to populations of broken and unbroken hydrogen bonds. The shape of $\mathscr{P}(V)$ does suggest, however, that even though the distribution of hydrogen bonds per molecule is unimodal, there still can exist, by virtue of other interactions, a bimodal distribution of environments. Whether or not one chooses to call these environments hydrogen bonded and nonhydrogen bonded, or to use other adjectives, is a matter of motivation and taste.

For the particular case of simulated water based on the BNS potential the difference between energy centroids of the bimodal $\mathscr{P}(V)$ is 2.54 kcal/mole. For the ST2-potential this changes to 2.9 kcal/mole. This average energy difference between environments is in modest agreement with the value assigned in the bond excitation model proposed by Angell [67] (1.9 kcal/mole).

Further information about the topology of the hydrogen bonding is obtained from the distribution of polygons without cross bonding (non short circuited). These are shown in Figs. 31–34 and the results are also set out in Table 9. If the topology of the liquid bonding were identical with that of ice Ih, or ice Ic, except for random hydrogen bond breakage, only polygons with an even number of sides would occur. Clearly, this is not the case.

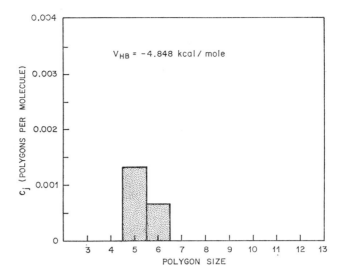

Fig. 31. Distribution functions for non-chort-sircuited polygons as predicted by molecular dynamics simulation of liquid water (from Ref. [3])

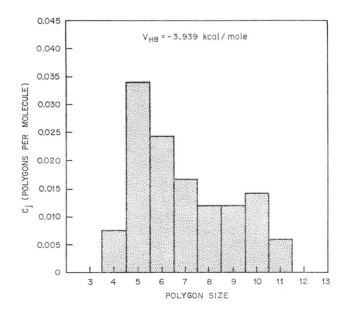

Fig. 32. Distribution functions for non-short-circuited polygons as predicted by molecular dynamics simulation of liquid water (from Ref. [3])

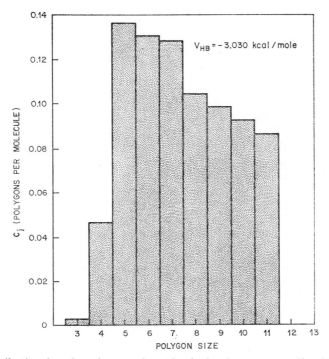

Fig. 33. Distribution functions for non-short-circuited polygons as predicted by molecular dynamics simulation of liquid water (from Ref. [3])

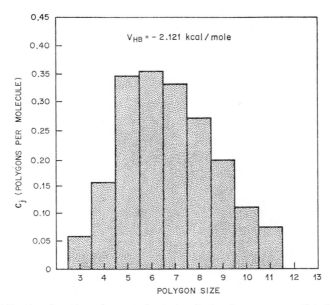

Fig. 34. Distribution functions for non-short-circuited polygons as predicted by molecular dynamics simulation of liquid water (from Ref. [3])

Table 9. Parameters characterizing the hydrogen-bond patterns in liquid water at 10 °C, and mass density 1g/cm³. The mean number of hydrogen bonds terminating at a molecule is $\langle b \rangle$; n_0 is the fraction of unbonded molecules, and n_1 is the fraction with precisely one bond. C_j stands for the number of non-short-circuited polygons, per molecule of the liquid, with j sides

	I	II	III	IV
V_{HB} (kcal/mole)	—2.121	—3.030	—3.939	—4.848
$\langle b \rangle$	3.88	3.14	2.26	1.18
n_0	0	0.00331	0.0410	0.249
n_1	0.0026	0.029	0.180	0.415
C_3	0.05952	0.002976	0	0
C_4	0.1564	0.04663	0.007606	0
C_5	0.3459	0.1362	0.03406	0.001323
C_6	0.3548	0.1306	0.02447	0.0006614
C_7	0.3320	0.1280	0.01687	0
C_8	0.2715	0.1045	0.01224	0
C_9	0.1971	0.09854	0.01224	0
C_{10}	0.1118	0.09292	0.01422	0
C_{11}	0.07573	0.08664	0.005952	0

Finally a decomposition of $g_{OO}(R)$ into angular components according to

$$g_{OO}(R) = g_{OO}^{Tet}(R) + g_{OO}^{Res}(R) \qquad (5.4)$$

where $g_{OO}^{tet}(R)$ describes correlations between molecules in the cone of angles defined by the tetrahedrally opposed faces of an octahedron (see Fig. 35), and $g_{OO}^{res}(R)$ the correlations in the remaining portion of angle space, confirms that

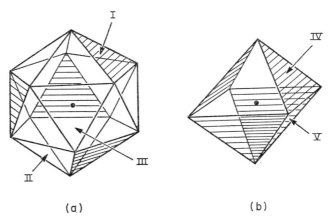

(a) (b)

Fig. 35. Polyhedral bases for orientational resolution of g_{OO}. The four shaded faces for both the icosahedral (a) and octahedral (b) schemes are those whose centers may simultaneously be pierced by a tetrahedral set of directions (*i.e.* undistorted H-bonds). The Roman numerals identify face classes (see Fig. 36). (From Ref.[13])

the dominant feature of the local structure is tetrahedral ordering of nearest neighbor oxygens (see Fig. 36). Notice, however, that deviation from tetrahedrality, in particular penetration into the vicinity of the central oxygen, also occurs.

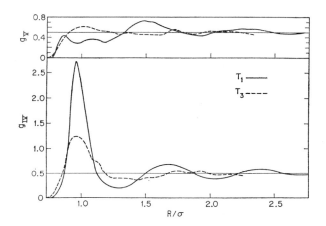

Fig. 36. The tetrahedral and residual components of the OO pair correlation function as predicted by molecular dynamics simulation of liquid water (from Ref. [3]). Note that $T_3 > T_1$. For the unscaled BNS potential $T_1 = 265$ K, $T_3 = 588$ K.

Because simulated water is a classical liquid, the computed power spectrum which describes the translational motions, is bound to disagree with that of real water. Figure 37, shows that the power spectrum has peaks at ~ 44 cm^{-1} and ~ 215 cm^{-1}, whereas for real water they occur at ~ 60 cm^{-1} and 170 cm^{-1}. A similar discrepancy exists between simulated and real water rotational power spectra (compare the simulated water frequencies 410 cm^{-1}, 450 cm^{-1} and 800–925 cm^{-1} with the accepted experimental values 439 cm^{-1}, 538 cm^{-1} and 717 cm^{-1}). In this model localization of the molecules around their momentary orientations is only marginal.

Stillinger and Rahman have also considered the diffusion coefficient, velocity autocorrelation function and scattering function for simulated water. For discussion of these interesting calculations the reader is referred to their papers [3].

To sum up, the molecular dynamics and Monte-Carlo simulations of water lead to the conclusions:

(i) The local environment of a water molecule is dominantly tetrahedral,

(ii) There is no evidence for presence in the liquid of ice microcrystallites, or of other species with specific geometries,

(iii) The distribution of hydrogen bonds per molecule is unimodal, but there may exist a bimodal distribution of environments with respect to average interaction energy,

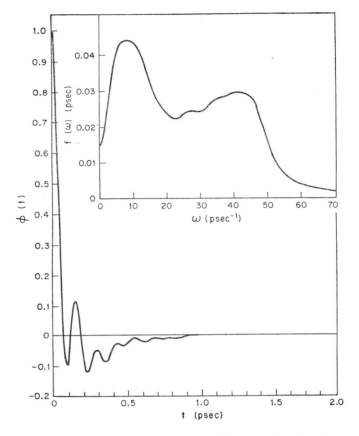

Fig. 37. The translational power spectrum as predicted by molecular dynamics simulation of liquid water (from Ref. [3])

(iv) The simplest visualization of the structure, beyond the local tetrahedrality of environment of each molecule, is that there is a randomized, imperfect, but space filling network of hydrogen bonded molecules; some of the hydrogen bonds are severely bent, as shown in the Monte-Carlo configurations displayed in Figs. 38 and 39.

Note added in proof. The application of the usual integral equation theories of the liquid state [2] to water has not been successful.[1] A recent study by H. C. Andersen [J. Chem. Phys. 61, 4985 (1974)] promises to change this situation. Briefly, Andersen reformulates the well known Mayer cluster expansion of the distribution function [2] by consistently taking into account the saturation of interaction characteristic of hydrogen bonding. Approximations are selected which satisfy this saturation condition at each step of the analysis. Preliminary calculations (H. C. Andersen, private communication) indicate that even low order approximations that preserve the saturation condition lead to qualitative be-

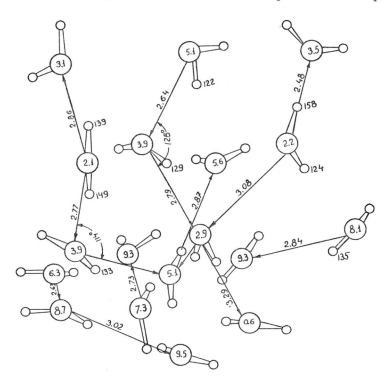

Fig. 38. A typical molecular configuration from the Monte Carlo simulation of liquid water (from Ref. [71]). Hydrogen bond lengths and OH...O angles are displayed. The numbers inside the oxygen atom circles are values of the Z coordinate in Å

havior (*e.g.* density maximum, about four nearest neighbors, etc.) like that observed in water.

Another recent development is the study of a central force model of liquid water by the method of molecular dynamics (A. Rahman, F. H. Stillinger and H. Lemberg, private communication). In this model the molecule-molecule interaction, such as represented by the BNS potential, is replaced by a sum of atom-atom potentials for all atoms in the system. The individual atom-atom potential forms are chosen such that (a) $H+H+O$ triads spontaneously settle into the geometry of free H_2O molecules and (b) the net interaction between two H_2O triads leads to formation of a linear hydrogen bond. In the absence of distortion of the participating H_2O molecules, the hydrogen bond has energy 6.12 kcal/mole at an OO separation of 2.86 Å. The intramolecular vibrational frequencies of H_2O are well reproduced ($\nu_1^{\text{expt}} = 3657$ cm^{-1}, $\nu_1^{\text{theory}} = 4266$ cm^{-1}, $\nu_2^{\text{expt}} = 1595$ cm^{-1}, $\nu_2^{\text{theory}} = 1369$ cm^{-1}, $\nu_3^{\text{expt}} = 3756$ cm^{-1}, $\nu_3^{\text{theory}} = 3805$ cm^{-1}).

The results of calculations based on the central force model are very satisfying. As shown in Figs. C, D and E the pair correlation functions $g_{\text{HH}}(R)$, $g_{\text{OH}}(R)$

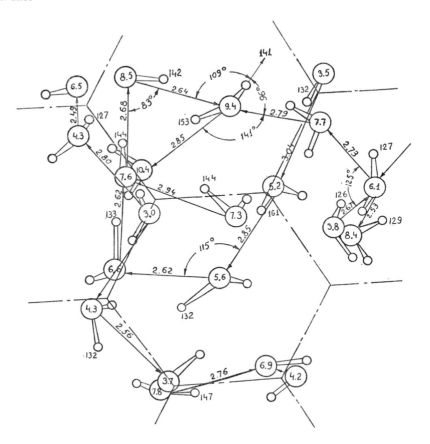

Fig. 39. A typical molecular configuration from the Monte Carlo simulation of liquid water (from Ref. [71]). Hydrogen bond lengths and OH...O angles are displayed. The numbers inside the oxygen atom circles are values of the Z coordinate in Å

and $g_{OO}(R)$ have just the expected properties. The molecules in the sample all remain intact, and each has exactly the correct number of H's and O. The curve for $g_{OO}(R)$ shows 4.5 neighbors at 2.86 Å, and the ratio of second to first neighbor distances is 1.57, both in tolerably good agreement with experimental data. In general, the existence of linear hydrogen bonding between near neighbors is supported by the computational results.

A very sensitive test of the model is the comparison of calculated and observed structure functions; this is shown in Figs. F and G. Note that the central force model yields the characteristic double peak in $s\tilde{h}(s)$ near 2.5 Å^{-1}. That the theoretical curve oscillates with greater amplitude than the experimental data indicates that the predicted distribution of near neighbor OO separations is too narrow. The comparison with neutron diffraction data shows that the theoretical maximum in $\tilde{H}(s)$ is too small, and the fall off as $s \to 0$ is incorrect, the latter probably being an artifact of calculational procedure rather than of the model.

In general, this central force approach appears to offer valuable insights into the structure of water, including dynamical effects. It should be improved and studied further.

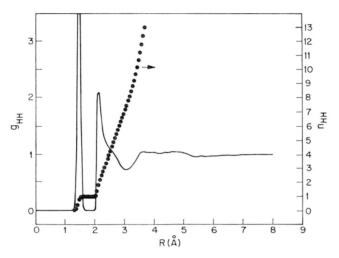

Fig. C. The pair correlation function g_{HH} and the running coordination number n_{HH},

$$n_{\alpha\beta} = 4\,\pi\,\varrho_\beta \int_0^{R'} R^2\,g_{\alpha\beta}(R)\,dR\,,$$

predicted by the central force model

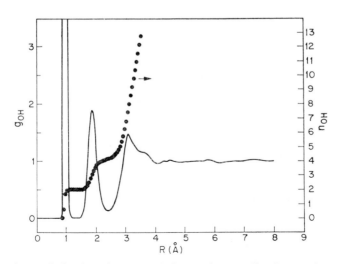

Fig. D. The pair correlation function g_{OH} and the running coordination number n_{OH} predicted by the central force model

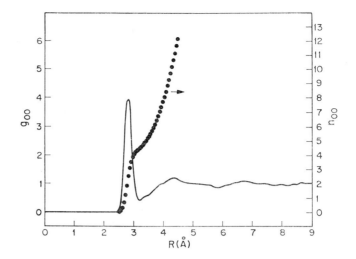

Fig. E. The pair correlation function g_{OO} and the running coordination number n_{OO} predicted by the control force model

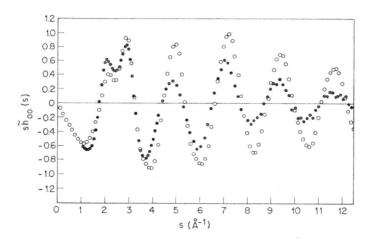

Fig. F. The structure function for X-ray diffraction predicted by the central force model O, and the experimental data of Narten ⬤

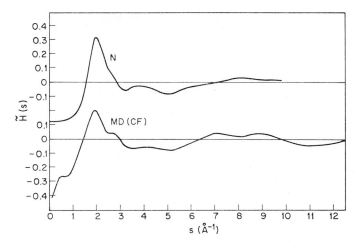

Fig. G. The structure function for neutron diffraction predicted by the central force model (MDCF) and the experimental data of Narten (N)

VI. Conjectures Concerning the Structures of Amorphous Solid and Liquid Water

In this section we attempt to piece together, self-consistently, the robust information contained in the results described earlier. For this purpose we regard the predictions of the several theoretical models as hints to be selected and followed up in conjunction with experimental data. Our goal, of course, is the construction of a conjecture concerning the structures of amorphous solid and liquid water.

The most direct source of structural information, the diffraction studies, provides strong evidence for predominantly tetrahedral ordering of a molecule and its nearest neighbors, both in amorphous solid and liquid H_2O. Other, weaker but still direct, structural evidence comes from the ratio of separations of nearest neighbor and next nearest neighbor OO pairs in liquid water, the ubiquity of tetrahedral ordering in the several crystalline ices, and the statistical geometry of simulated water.

Useful, but less direct, information about the local ordering in H_2O(as) and liquid H_2O comes from studies of the hindered translation and libration regions of the vibrational spectrum. It is worthwhile examining what is specific, and what general, in these studies. The existence of peaks in the density of vibrational states near 60 cm^{-1} and 170 cm^{-1} is predicted by all of the models described in Section V. These models, ranging from simulated water (which employs no initial structural information), to the cluster model (which uses as input the hypothesized structures of a large number of species) all employ a strongly tetrahedral water-water interaction. An even simpler model, namely a water pentamer with C_{2v} symmetry (see Fig. 40), has been invoked by Walrafen [39] to explain the 60 cm^{-1} and 170 cm^{-1} peaks. Of course, bond length and OOO angle distributions with non-zero widths are characteristic of the local environment of a water molecule in the liquid, and the mentioned symmetry assignment is intended to be understood as referring to an average structure. In terms of this average structure, treat-

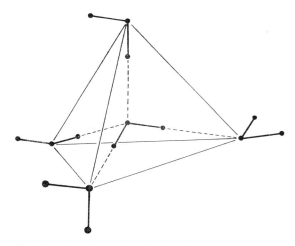

Fig. 40. A local configuration showing C_{2v} pentamer symmetry

ing the pentamer as a molecule and neglecting the interactions with the rest of the liquid, there are nine intermolecular vibrations, of which six refer to hindered translations and three to restricted rotations. Two of the six hindered translations are equivalent to hydrogen bond bending motions [$\nu_3(A_1)$ and $\nu_4(A_1)$] and the remainder to hydrogen bond stretching motions [$\nu_1(A_1)$, $\nu_2(A_1)$, $\nu_7(B_1)$ and $\nu_9(B_2)$]. Walrafen assigns the 60 cm^{-1} peak in the liquid water spectrum to (unresolved) ν_3 and ν_4, and the 170 cm^{-1} peak to (unresolved) ν_1, ν_2, ν_7 and ν_9.

The preceding suggests that the structure of the density of vibrational states in the hindered translation region is primarily sensitive to local topology, and not to other details of either structure or interaction. This is indeed the case. Weare and Alben [35] have shown that the density of vibrational states of an exactly tetrahedral solid with zero bond-bending force constant is particularly simple. The theorem states that the density of vibrational states expressed as a function of $M\omega^2$ (in our case M is the mass of a water molecule) consists of three parts, each of which contains one state per molecule. These are a delta function at zero, a delta function at 8α, where α is the bond stretching force constant, and a continuous band which has the same density of states as the "one band" Hamiltonian

$$\mathcal{H} = \alpha \left(4\,1 - \sum_{l\Delta} |l> <l\,\Delta| \right). \tag{6.1}$$

In Eq. (6.1) 1 is the unit operator, there is one state $|l>$ associated with each lattice site, and l and $l\Delta$ ($\Delta = 1, 2, 3, 4$) label a molecule and its nearest neighbors in the tetrahedral lattice. Weare and Alben show also that the theorem remains valid when small distortions away from tetrahedrality exist, hence it can be used to describe a random amorphous solid derived from a tetrahedral parent lattice. Basically, the density of states of the amorphous solid is a somewhat washed out version of that of the parent lattice. The general shape of the frequency spectrum is not much altered by the inclusion of a non zero bond-bending force constant provided the ratio of it to the bond stretching force constant is small relative to unity.

To apply the Weare-Alben theorem to water we start with the observation that the relative positions of maxima in the densities of vibrational states of crystalline Si, Ge and ice Ic are similar (see Table 10). Next, we display in Fig. 41 the

Table 10. Relative positions of maxima in the densities of vibrational states of crystalline Si, Ge and ice Ic

Mode	Ice Ic	Si	Ge
TO[1]	1.000	1.000	1.000
LO[2]	0.829	0.858	0.857
LA[3]	0.716	0.692	0.647
TA[4]	<0.220	0.264	0.232

[1] Transverse optic mode.
[2] Longitudinal optic mode.
[3] Longitudinal accoustic mode.
[4] Transverse accoustic mode.

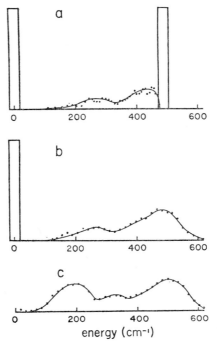

Fig. 41. Density of states for a continuous random tetrahedral network model (a) one band Hamiltonian [Eq. (6.1)], (b) $\beta = 0$, (c) $\beta = 0.2\,\alpha$. In each case $\alpha = 0.495 \times 10^5$ dyne/cm. (from Ref. [35])

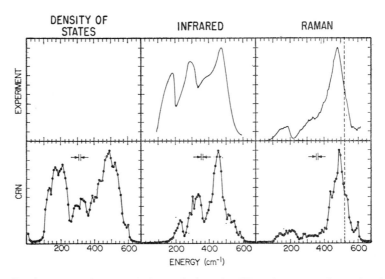

Fig. 42a. Continuous random network predictions for IR and Raman absorption in Si(as) and experimental data (from Prof. R. Alben)

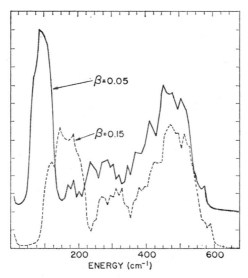

Fig. 42 b. Density of states for a continuous random network for $\beta/\alpha = 0.05$ and $\beta/\alpha = 0.15$ with $\alpha = 0.475 \times 10^5$ dyne/cm. (from Ref. [35])

phonon density of states for a continuous random network model with tetrahedral parentage, and in Fig. 42, a comparison of observed and computed infrared and Raman spectra for amorphous Si. It is clear that what one observes in a given spectrum depends on details of the breakdown of free molecule and perfect crystal selection rules. Nevertheless, the prominent feature of these results that is of use to us is that the ratio of frequencies of the two largest peaks in the density of vibrational states of Si(as) is ~3 (ratio of bending to stretching force constants of 0.15), and of the two peaks in the Raman spectrum is ~2.5. The ratio of frequencies of the two peaks in the hindered translation region of the Raman spectrum of liquid H_2O is nearly the same, namely ~2.8.

We conclude, by inverting the argument just presented and knowing the structure of Si(as), that the low frequency region of the spectrum of liquid water is in complete agreement with the existence of local tetrahedral ordering, that this ordering can be understood to represent an average over a range of OOO angles, and that neither a specific molecular species (e. g. only a pentamer) nor long range order are required to interpret the observations.

A similar but less forceful argument can be made with respect to interpretation of the librational region of the spectra of H_2O(as) and liquid H_2O.

The most difficult to interpret of the available evidence is the nature of the vibrational spectra of the OH stretching regions of H_2O(as) and liquid H_2O. This particular subject is a minefield littered with conflicting opinions and sometimes emotionally held positions. I do not pretend that the following discussion is definitive, but I think it is consistent with the other information available to us.

The decomposition of the observed Raman (or infrared) spectral contours, in particular the positions and assignments of the component bands, can be used as the basis for conjecture concerning the local environment of a water molecule. We start with the data for (low temperature) H_2O(as), D_2O(as) and HDO/D_2O(as).

If the assignments suggested in Section IV are accepted the obvious first conclusion is that molecules experience, effectively, three different environments. We repeat our earlier admonition: this should not be taken to imply the existence of three species. Rather, our view is that the distribution of environments, as defined by the criterion of OH stretching frequency, has three incompletely separated peaks. These we visualize as corresponding to strong almost linear hydrogen bonding, intermediate strength bent hydrogen bonding, and weak (probably many configurations) hydrogen bonding. The fact that there are only two components in the decomposed Raman spectra of HDO/D_2O(as) can be accomodated in this picture if the decoupled OH stretching motions of molecules with both nearly linear and bent hydrogen bonds have frequencies close enough not to be resolved. The observation that the component band widths for HDO/D_2O(as) are $\sim 50\%$ larger than the corresponding band widths for H_2O(as) is consistent with this suggestion.

As already argued, the most probable configuration of a molecule and its first neighbors involves strong symmetric hydrogen bonding. We identify the mode ν_1 of such a central molecule with the intense first component in the OH stretching region of the spectrum of H_2O(as). Note that the frequency of this band (ν_1^{sb}) is 3110 cm^{-1} in H_2O(as), which is close to the value of ν_1 in ices Ic and Ih, namely 3085 cm^{-1} (at 77 K) [1]. That this similarity in stretching frequencies should exist is suggested by the near equality of nearest neighbor OO separations in these phases: 2.76 Å [H_2O(as) at 10 and 77 K], 2.755 Å (ices Ic and Ih at 93 K) [1]. On the other hand, the vibrational band width in ice Ih is smaller (~ 40–50 cm^{-1}) [1] than in H_2O(as) (~ 80 cm^{-1}) [8]. At least some part of the increased breadth can be accounted for by recognizing that the distribution of OOO angles in ice Ih is much narrower than in H_2O(as). Clearly, by virtue of the distribution of interactions which follows from a distribution of OOO angles, one expects a broadening of the distribution of possible values of ν_1^{sb} (and also ν_3^{sb}).

We have already pointed out that the breadth of ν_1^{sb} in HDO/D_2O(as) is greater (~ 115 cm^{-1}) than that of ν_1^{sb} in H_2O(as). In addition to the proposed overlapping of the bands ν_1^{sb} (HOD) and ν_1^{bb} (HOD), we must allow that any difference in hydrogen bonding character between OH...O and OD...O will also contribute to the breadth of the distribution of local environments, hence also to the breadth of the transition. Presumably any such contribution to variation in the local environments is in addition to the effects already present in H_2O(as). Of course, at the temperatures used by VRB, thermal broadening is negligibly small relative to the broadening from the other sources mentioned.

Having called attention to the similarity in the values of ν_1 for H_2O(as) and ice Ih, we must now call attention to the difference. In the case of the fully coupled OH stretching motion this is ~ 25 cm^{-1}; in the case of the uncoupled OH stretching motion this is also ~ 25 cm^{-1} in the same direction [ν_1 H_2O(as) $> \nu_1$ (ice Ih)]. It is interesting that the uncoupled value of ν_1 in ice II is 3373, 3357 and 3323 cm^{-1}, in ice III is 3318 cm^{-1}, in ice V is 3350 cm^{-1} and in ice VI is 3338 cm^{-1} [27]. Each of these ice forms has a more complex crystal structure than ice Ih. In general ices II–VIII have higher density than ice Ih, and have some severely bent hydrogen bonds [1,11]. Given the lack of thermal excitation in the low temperature samples studied by VRB, the demonstrated lack of temperature dependence of the deduced intramolecular frequencies, and the direction of the deviation of the values of ν_1 from the ice Ih value, it is tempting to infer that at least some of the water molecules in low temperature H_2O(as) have local environments which are distorted variants of lattices more complex than that of ice Ih.

To proceed further we require some information about the structures of ices Ih, II and III, and how these structures can be converted, one to the other. In the following we draw heavily upon an excellent discussion published by von Hippel and Farrell [74].

Note added in proof. The following description of the conversions ice Ih → ice II, etc. is not intended to represent the actual mechanism of the physical process. It is meant to be only a visualization of the geometric changes which can be used to convert one lattice into another. Clearly, the real mechanism of conversion depends upon the nature of the many molecule potential energy surface. Possible pathways for interconversion of the several crystalline ices are under investigation by R. McGraw and S. A. Rice. Perhaps the most interesting of their preliminary results is the finding that it is rather easy to build an ice III lattice contiguous to an ice Ih lattice with only slight boundary distortion, and that, what are in essence proton transfer generated inversions, something like pseudo rotations, can be used to effect the conversion insofar as symmetry is concerned. At present the study is incomplete, and the energy changes along the several pathways possible are not yet calculated, hence there is no basis for choosing between them.

Consider first ice Ih (Figs. 43 and 44), which is here visualized as an array of hexagons of both "boat" and "chair" forms. Two parallel puckered rings, rotated by 60° and then hydrogen bonded, form a cage defining the (threefold) c axis of ice Ih normal to the ring plane, and the three a axes in that plane. The three sides of this cage are boat shaped rings.

Because the work required to bend hydrogen bonds is much less than that required to stretch (or compress) them, it is easier to generate different crystalline ices by distorting OOO angles at constant OO separation than by changing the OO spacing. Ice II can be generated from ice Ih by breaking hydrogen bonds in the Ih hexagonal rings in favor of new hydrogen bonds formed diagonally across the hexagonal "channels" that are parallel to the c axis of ice Ih (see Fig. 44). Viewed in the puckered plane perpendicular to the ice Ih c axis, by connecting and breaking the hydrogen bonds as shown in Fig. 45 one half the hexagon rings are destroyed, the surviving columns of hexagon rings are elevated or depressed relative to their original locations, and the old hexagonal plane is differentiated into strongly puckered (O''_I) and almost flat (O''_{II}) rings. Each O''_I ring is laterally connected to six O''_{II} rings, and vice versa. Whereas in ice Ih there is proton disorder because of the existence of six equivalent OOO tetrahedral angles, the formation of ice II by the diagonal hydrogen bonding described increases four of the OOO angles to large values, whilst two are closed to near 90° and become the preferred locations of protons. The differentiation of O''_I from O''_{II} rings follows from proton ordering (Fig. 46).

While ice Ih → ice II by hydrogen bonding diagonally across the hexagons of the puckered planes perpendicular to the Ih c axis, ice Ih → ice III by hydrogen

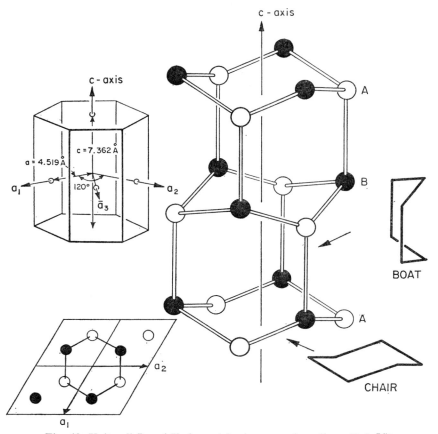

Fig. 43. Unit cell Ice of Ih formed by hexagon rings (from Ref. [74])

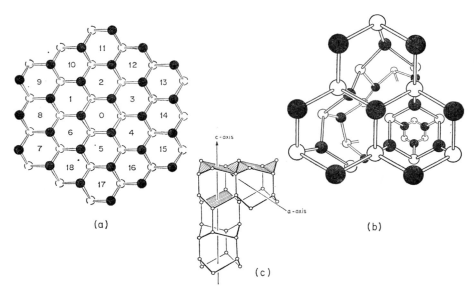

Fig. 44. Hexagon structure of Ice Ih: (a) reference plane of puckered rings perpendicular to
"c"; (b) hexagon channels viewed parallel to "c"; (c) fore-shortened channel in "a" direction
(from Ref. [74])

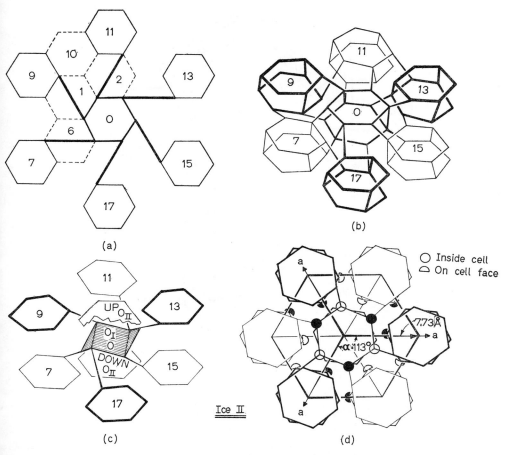

Fig. 45. Formation of Ice II: (a) Hexagon plane of ice Ih with diagonal bond connections; (b) up-down displacement of surviving hexagon columns; (c) O''_I hex ring (strongly puckered) connected to 8 O''_{II} rings (almost flat); (d) unit cell of ice II (from Ref. [74])

bonding to second nearest neighbors in puckered (chair) rings and in boat rings (see Fig. 46). Moreover, in the formation of ice II one half of the original hexagon rings of ice Ih are destroyed; in the formation of ice III all of the original hexagon rings of ice Ih are destroyed and replaced with pentagon rings. Three of the five OO bonds of a pentagon belong to three rings, and the other two to only one ring. Just as in ice II there is a differentiation of H_2O molecules into two types, O'''_I and O'''_{II}. Aside from remarking that each pentagon consists of three O'''_I and two O'''_{II} type molecules, we will not pursue further the complicated bonding relationships.

We note that in the ice II and ice III structures (and in all others except ice VII) each water molecule has four neighbors at ~2.75 Å, followed by a number of more distant neighbors. These are near 3.24 Å in ices II and III [1,11,75]. In ice Ih the nearest non hydrogen bonded neighbors are at 4.50 Å. The increase in the

179

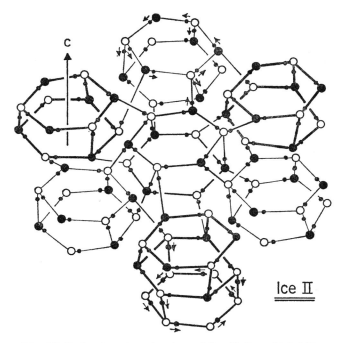

Fig. 46. Ordered proton structure of Ice II (from Ref. [74])

higher ices in number of non nearest neighbors is accomplished by bending hydrogen bonds. Bending distortion is, however, minimized by a "doubling back" in the bond network, so that what are real space non hydrogen bonded near neighbors are, in the network topology, more distant neighbors. For example, the 3.24 Å neighbor in ice II is topologically (along the network) a third neighbor. In ices Ih, II and III there is a single complete tetrahedral network, but in ices II and III as already stated, the OOO angles are distorted from the perfect tetrahedral value.

We now return to the analysis of the decomposition of the H_2O(as) vibrational spectrum in the OH stretching region. It should be made clear to the reader that what follows is highly speculative; the "results" will be of use, however.

Noting that the nearest neighbor OO separation in ices I, II and III is the same as that in H_2O(as), a crude correlation between OH stretching frequency and hydrogen bond angle can be formulated unter the assumption that the higher the frequency the greater the deviation of the hydrogen bond from linearity. The suggested correlation is shown in Fig. 48. The assigned values for ν_1^{sb} H_2O(as) and ν_1^{sb} D_2O(as) correspond to hydrogen bond donor angles of ∼107°, and the full width at half maximum of the band contour corresponds to an angular spread of ∼6°. The assigned values for ν_1^{bb} correspond to hydrogen bond donor angles ∼97°, close to one of the donor angles of both ice II and ice III [1,75]. If the peaks ∼3370 cm^{-1} H_2O(as) and ∼2478 cm^{-1} D_2O(as) are, instead, assigned to ν_1^{bb}, the corresponding donor angles are ∼87°, again close to one of the donor angles

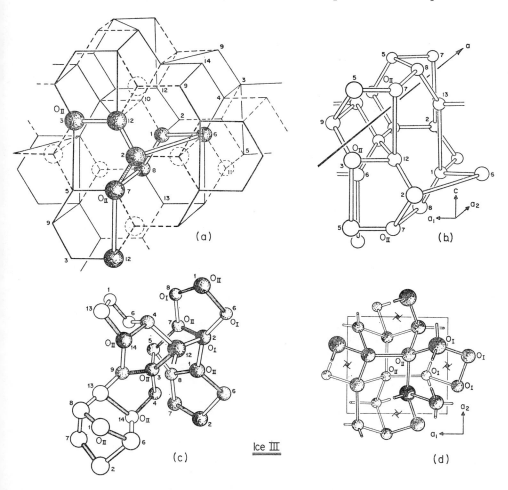

Fig. 47. Formation of Ice III: (a) Second neighbor connections across puckered rings and boat raings; positions of discarded H_2O molecules; (b) pentagon rings and channel in "a" direction; (c) O_I''' and O_{II}''' lattice points sharing one and two O—O bonds respectively with three pentagons; (d) O_I''' columns defining fourfold screw axis (from Ref. [74])

of both ice II and ice III. In either case the full width at half maximum of the band contour corresponds to an agular spread $\sim 12°$. Finally, the assigned values for v_1^{wb} lie off the scale defined by the correlation, hence presumably correspond to donor angles so far from $109°$ that the resultant hydrogen bonding is very weak.

The preceding suggestion of environments with ice II-like and ice III-like character in low temperature $H_2O(as)$ is consistent with its density (1.1 gm cm^{-3}; that of ice II is 1.18 gm cm^{-3} at 238 K, and of ice III is 1.15 gm cm^{-3} at 251 K), and the existence of the 3.3 Å peak in the OO distribution function (both ice II and ice III have an OO separation of ~ 3.24 Å). Furthermore, if we regard ice Ih as the prototype strongly hydrogen bonded system, we have (from infrared data)

181

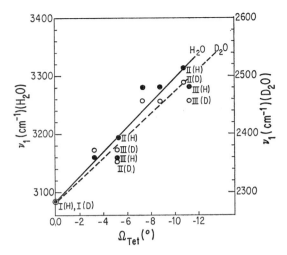

Fig. 48. The correlation between OH stretching frequency and H bond donor angle in crystalline ices I, II, III. $\Omega_{tet} = \frac{1}{2}(\theta_{donor} - 109.5)$ (from Ref. [8])

the relationship ν_1^{sb} (ice Ih) $< \nu_1^{sb}$ H_2O(as) $< \nu_1$ (ice II, ice III), clearly the order necessary to be consistent with the proposed mixture of environment in H_2O(as).

We are, therefore, led to propose that low temperatures H_2O(as) has a random network structure of mixed lattice parentage, i.e. that the structure is characterized by locally random OOO angle deviations from ice Ih (or Ic), ice II and ice III type lattices.

Since each of these lattices has a single hydrogen bonded network, and bond switching can change one to the other (with some volume change for Ih → II and Ih → III, but none for II → III), the topology of such a mixed random network should not be too difficult to generate. We cannot, at present, estimate the fractions of randomized environments of the several types. Given the density, one might guess the ice I-like environment to be rare relative to those with ice II or ice III parentage.

Note added in proof. Earlier in the text it was mentioned that the model used to describe the structure function of low density H_2O(as) does not describe that of high density H_2O(as). However, Narten, Venkatesh and Rice [27] do show than an ice I-like network with a near neighbor distance of 2.76 Å has the density and distance spectrum of high density H_2O(as) *if* one permits 45% of the cavities characteristic of this structure to be occupied by water molecules. These are not ordinary "unbonded" interstitials. If the cavity molecules are located on the c axis at a distance of 2.76 Å from the nearest network molecule each cavity molecule would have second neighbor network molecules at a distance of 3.25 Å. Moreover, since occupancy of 45% of the cavities implies that 81% of the water molecules are part of the tetrahedral network and 19% in cavity positions, the average coordination number of nearest neighbors in this model is 4.3, as is found for H_2O(as) 10 K/10 K. Structure functions calculated for this interstitial variant of a randomized ice I model (the randomization is effected as in the simple ice I model) are in good agreement with the observed structure function of $H_2O(as)$ 10 K/10 K derived from the X-ray data.

How, if at all, is this model related to the one implied in the discussion of ice II and ice III — like local environments? Note that the two models mentioned have an important common property, namely the presence of OOO angles which deviate considerably from 109.5° and which appear as distinct peaks in the OOO population versus angle distribution function. Now, placing an H_2O molecule in an interstitial position in an ice I lattice also leads to the introduction of unusually small OOO angles ($\sim$70°). We think it likely that any model of high density $H_2O(as)$ must include OOO angles much smaller than 109.5°.

The very suggestion of two models implies that neither is unique. Nor does the citing of two models exhaust the structural possibilities. Suppose it is possible to construct several more models of high density $H_2O(as)$, each derived from some crystalline prototype. The distinctions between these structures and an ice I structure with interstitial molecules are important and justified for the crystalline parents, but it is not clear whether such distinctions are meaningful for the amorphous phase of solid water because the local environments of water molecules are severely distorted from the average configuration, these distortions leading to the complete loss of all positional correlation a few molecular radii away from any starting point. In view of these distortions, distinction between random network models of unique parentage (Ice Ih with interstitials) and of more complex parentage (high pressure ice polymorphs) will be meaningful only if there exist observable properties assignable to some property of the parent lattice (e.g. proton ordering as in Ice II).

Following the same line of thought, we are led to propose that high temperature $H_2O(as)$ has a simple random network structure derived from an ice I type lattice [i.e. like Ge(as) and Si(as)].

Last, and most speculative, we propose that liquid H_2O also has a random network structure.

In this case the high molecular mobility and the extensive thermal excitation should lead to distributions of OOO angles and OO separations with much larger dispersions than their $H_2O(as)$ counterparts. Given the interconvertability of the ice Ih, ice Ic, ice II and ice III lattices, each of which is a single connected network, we expect there to be a range of network preserving local environments that resemble distortions of all these lattice types. The interconversions are governed by a constraint, namely that the macroscopic density, which is intermediate between that of ice I and ice II or ice III densities, remain fixed. The different local environments are, in this view, generated continuously by fluctuations.

The idea underlying the structural hypotheses advanced is that the many molecule potential energy surface in water has minima corresponding to many geometries, and that it takes but slight cooperative repositioning and reorienting of a group of molecules to completely alter the positions and depths of these minima. That this is so is only partly tautology: direct study [76] of model water-water pair interactions shows that a different pair geometry is stable for OO separation less than a critical distance R_c (R_c = 4.964 Å for the ST-2 potential) than for OO separation greater than this distance (see Fig. 49). For small distances

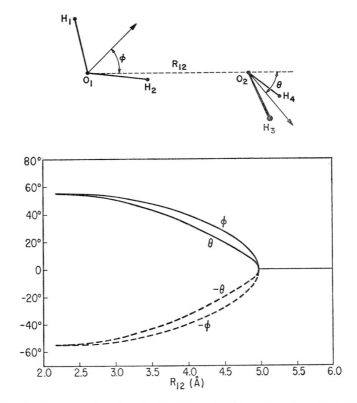

Fig. 49. Symmetry breaking in the ST_2 potential surface (from Ref. [76])

$(R < R_c)$ the linear hydrogen bonding is most important, while for large distances $(R > R_c)$ the dipole-dipole interaction promotes alignment of the molecular axes. At R_c eight equivalent small distance minima of low symmetry and four equivalent large distance minima of high symmetry have the same energy. Although the symmetry properties of the many molecule energy surface are undoubtedly more complicated than this, residues of such behavior are to be expected for that surface.

We have constructed the several structural hypotheses mentioned above so as to be in agreement with available experimental data (or at least our interpretation of them). Testing of these hypotheses requires new experimental data (see Section VII). For the present we turn to an examination of the consistency of the ideas advanced.

For reasons which will become clear, we examine first the case of high temperature H_2O(as). Two random network models relevant to our hypothesis have been described in the literature. Both are based on distortions from a single locally tetrahedral structure that is like ice Ih. Kell's model [77] is much too small to be very useful. Nevertheless, its successful construction, just as for the case of Ge(as) [78], SiO_2(as) [79], and others, shows the viability of the random network concept.

namely that it *is* possible to build a fully connected structure with modest deviations from the classical tetrahedral angle and without long range order. In Kell's model the root mean square deviation from 109° is 8.4°, the mean deviation being 6°. By comparison with Pople's random network model [80] of the dielectric constant of liquid water, it is estimated that Kell's model corresponds to a structure at $T \approx 150$ K. The model can be used to predict that the coefficient of thermal expansion of H_2O(as) below ~150 K should be almost the same as that of ice Ih, and that if it could be maintained at higher temperature the amorphous solid would have a negative coefficient of thermal expansion. Disregarding quantitative details, the last prediction is striking when compared with the observed behavior of supercooled water [81].

Note added in proof. Kell has recently reported on another random network model of water [G. S. Kell, Chem. Phys. Lett. *30*, 223 (1975)]. In this model there is a random tetrahedral network built around linked eight member rings. It is found that the resulting "solid" is denser than a purely tetrahedral random network and that there are near neighbors in space derived from molecules which are not near neighbors along the bond network. The distribution of second neighbors is broad, corresponding to an interquartile angular spread of 26°. In Kell's previous tetrahedral random network [77], the interquartile angular spread corresponding to the second neighbor distribution is 12°.

A much more satisfactory random network model has been discussed by Alben and Boutron [82]. They used a model, proposed by Polk [78] for Ge(as), scaled to fit the observed nearest neighbor OO distance of H_2O(as), and with H atoms added to the OO bonds according to the "Pauling ice rule" that guarantees the presence of only H_2O molecules [65]. In the Polk model the bond length is everywhere the same and the OOO angles are distributed with root mean square deviation of 7° about 109°. For the case of Ge(as), the observed and model radial distribution functions are in excellent agreement.

Alben and Boutron compare their modified Polk model X-ray scattering structure function with that reported by Bondot and the model neutron scattering structure function with that reported by Wenzel, Linderstrøm-Lang and Rice. The fit to the X-ray data is generally good (see Fig. 50). It will be recalled that Bondot's sample of H_2O(as) was deposited at or above 77 K, hence should have a density of 0.93 gm cm^{-3}, corresponding (in this view) to a random network with parentage of only the ice Ih type. The Polk model is a random network of just this type, hence the agreement is consistent with the structural hypotheses advanced earlier. The agreement between the modified Polk model neutron scattering structure function and that observed is also generally good (see Fig. 51). Note that although the WLR sample was deposited at 7 K, it appears to be low density D_2O(as). It is disturbing that, at present, the conditions necessary for the generation at will of either high density or low density H_2O(as) are incompletely known. This uncertainty does not, however, interfere with construction of structural hypotheses and, with care, need not overly confuse testing those hypotheses.

Alben and Boutron suggest that the peak in the X-ray and neutron scattering functions at 1.7 Å^{-1} is indicative of an anisotropic layer structure extending over at least 15 Å in Polk type continuous random network models. To show this better Fig. 52 displays the radial distribution function of the Alben-Boutron modified

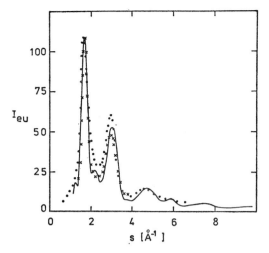

Fig. 50. Continuous random network fit to the X-ray diffraction pattern of $H_2O(as)$ (from Ref. [82]). Experimental data ● and ×; Theory ——

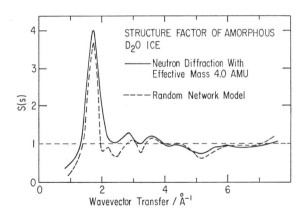

Fig. 51. Continuous random network fit to the neutron diffraction pattern of $H_2O(as)$ (from Ref. [82]). Experimental data ——, Theory ---

Polk model calculated for the case of neutron scattering and with the constituents having the isolated H_2O molecule shape and size. The oscillatory behavior for large R, with period ~3.7 Å, corresponds to the peak in the structure function at 1.7 Å^{-1}.

The reader should recall that the fitting of a structure to diffraction data is not unique. We have shown that both the "constructed" modified random network model of Polk, as well as the network simulated by allowing Gaussian distributions of atom-atom distances can fit the observed structure functions for low density $H_2O(as)$, and the latter, with modification to include small OOO

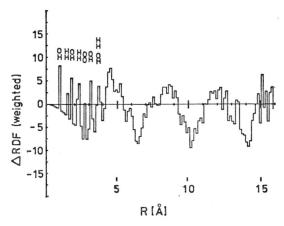

Fig. 52. The excess (above a smooth parabolic background) radial distribution function for $H_2O(as)$ weighted appropriately for neutron scattering from $D_2O(as)$ (from Ref.[82])

angles, can also fit the structure function of high density $H_2O(as)$. The general common features of these models are almost certainly correct representations of properties of $H_2O(as)$, but the detailed features of the models must be considered speculative and unsubstantiated.

Finally, we remark that the few thermodynamic data available are consistent with a simple continuous random network structure for high temperature $H_2O(as)$.

The two items that fall in this category are:

(i) The density of $H_2O(as)$ deposited at 77 K, measured by a flotation method[84], is 0.93 gm cm^{-3}, in agreement with the value inferred from X-ray data for a similarly prepared sample [7]. More important, it is characteristic of continuous random network models that the densities of amorphous and crystalline solids are nearly the same (to within 1—3%) [22]. The observed density is thus consistent with the assumed relationship between $H_2O(as)$ and ice Ih.

(ii) The specific heats [85] of $H_2O(as)$ and ice Ih are nearly the same below 135 K. A similar near equality of values of C_p for amorphous solid and crystalline phases in the low temperature region is found for Ge and other systems known to have simple random network structures [34]. At very low temperature C_p of an amorphous solid appreciably exceeds that of the corresponding crystal [e. g. $GeO_2(as)$, Se(as)] [34]. It seems likely this will also be true for $H_2O(as)$.

Consider, now, the hypothesized structure of the low temperature form of $H_2O(as)$. The key element in our considerations has been the relative ease of alteration of the many water molecule potential energy surface. That is, the evidence of the existence of many different crystalline ices, and the nature of their structures, suggests that a modest rearrangement of the positions of a group of water molecules can (and does) generate new minima in the potential energy surface, and that these new minima correspond to a qualitatively different connectivity of the hydrogen bond network. Then, given the exceedingly low molecular mobility at very low temperature, accidentally formed molecular arrangements, different from point to point in the deposit, can generate local minima in potential energy which belong to different networks of hydrogen bonds. There is every reason to believe that such local arrangements can resemble randomized versions of ices I, II, III, etc. At present we have no basis on which to pick amongst the several geometries, but it seems likely that the single network arrangements, which correspond to the most favorable condensed phase internal energies, are favored. It is for this reason that we speculate that low temperature $H_2O(as)$ is a mixture of randomized ice I, ice II and ice III types. Given the observed density of low temperature $H_2O(as)$, and the known densities and energies of ices II and III, it also seems likely they predominate in the mixed random network.

There do not yet exist random network models with mixed lattice parentage, hence the properties of low temperature $H_2O(as)$ cannot be compared with those predicted for such a model.

Turning, finally, to the case of liquid H_2O, we emphasize that the important element added to the already described random network model is the presence of molecular mobility, indeed high mobility. The consequences of this mobility are twofold. First, many local geometries are probed, hence many possible forms of connected (including badly distorted) hydrogen bond network are sampled. Second, because of the high mobility there are not any frozen configurations, and local arrangements with unfavorable potential energy, even if excited, are rapidly depopulated. The net effect is, in our picture, a continuous random network of mixed ice I, ice II, ice III, ... type parentage, but with much larger dispersion of OOO angles and OO separations than is characteristic of the immobile random network of an amorphous solid. We do not rule out the possibility that in the liquid the ice I-like component is much more important than in low temperature $H_2O(as)$. We do suggest that liquid H_2O differs from high temperature $H_2O(as)$ by virtue of the ice II-like, ice III-like, ... contributions. This is accompanied by an increase in density, in accord with the observation that the density of liquid H_2O is between those of ice Ih and ices II and III [hence also between those of high temperature and low temperature $H_2O(as)$]. In this view liquid H_2O is a very mobile version of low temperature $H_2O(as)$, with different fractional contributions from the several randomized lattice environments.

Calculations of the structure factor for the kind of random network proposed as a model of liquid water have not yet been made. It is possible, however, to guess some of the characteristics of the network. For example, given the suggested large dispersion of OOO angles and OO separations, it is reasonable to suppose that orientational correlation between molecules does not extend beyond the central water-nearest neighbors, tetrahedrally bonded group of molecules. Accordingly, a provisional test of the model can be made by comparing theoretical and experimental structure functions, where the theoretical function refers to a model with local tetrahedral bonding, no orientational correlation beyond nearest neighbors, partial penetration of the two nearest neighbor shell by molecules topologically distant (measured along the bonding network), and a continuum density outside the nearest neighbor tetrahedral structure. The results of these calculations, as reported by Narten [19], are displayed in Fig. 53. Clearly, the agreement between model and observed functions is very good. The deviation in the region 4.5 Å$\leq$

$R \leq 6$ Å is likely a consequence of inadequacy of the assumption that the density is uniform outside the central tetrahedral region.

The model just described does not conform in detail to the random network model proposed earlier. In particular, the use of a continuum outside the nearest-neighbor tetrahedral structure removes some of the correlations inherent in a continuous network. Nevertheless, given the existence of a broad OOO angular distribution, coupled to an OO distance distribution much broader than in H_2O(as), it is unlikely that this assumption introduces any features in serious disagreement with those characteristic of a random network model.

A different test, one less satisfactory because the standard of comparison is simulated water not real water, is obtained by examining the functions $h_{OD}(R)$ and $h_{DD}(R)$ predicted. The functions derived from the Narten model are shown in Fig. 54; they should be compared with those for simulated water, displayed in Figs. 27 and 28. Just as for the function $h_{OO}(R)$, the curves for simulated water are narrower and higher than those in Fig. 54. In all other respects the agreement between the two sets of functions is excellent. It now remains to be shown that a full calculation, based on precisely the random network model proposed will reproduce the data as well as this (sensibly equivalent?) model.

This is an appropriate point to remark on some of the thermodynamic aspects of the complicated random network structure envisaged for the liquid. Now, the thermodynamic properties of ices II and III are very similar [1]. The ice II → ice III transition (249 K, 3.4 kbar) involves only a very small change in volume, namely 0.26 cm³/mole, 1.6% of the molar volume), a small change in entropy 1.22 cal/° mole, and a small change in enthalpy, 304 cal/mole. Similarly, the ice I → ice II,

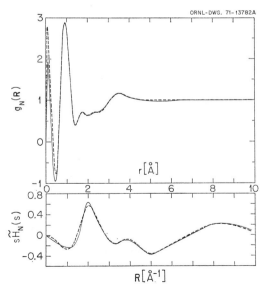

Fig. 53. Fit to the pair correlation function g_{OO} for the model consisting of a local pentamer plus continuum (from Ref. [5]). —— neutron diffraction data, --- model calculation

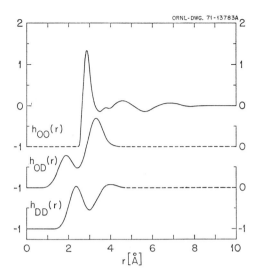

Fig. 54. Derived pair correlation functions from the data of Ref.[5]), using the model of a local pentamer plus continuum (from Ref.[5])

and ice I → ice III transitions involve entropy changes of only −0.76 cal/°mole and 0.4 cal/°mole respectively, coupled to enthalpy changes of only − 180 cal/°mole and 94 cal/mole, respectively. There are large (but almost equal) volume changes in the latter two transitions. It is not difficult to visualize local "randomized ice II" → "randomized ice III" changes. Furthermore, noting the difference in signs for the ice I → ice II and ice I → ice III transitions, it is possible to imagine local randomized configuration changes occuring in such a way that overall energy conservation is achieved with almost no net change in entropy, a change "randomized ice I" → "randomized ice II" in one locale being balanced by "randomized ice I" → "randomized ice III" somewhere else. In short, given nonzero molecular mobility and a reasonable range of allowable hydrogen bond distortions, the hypothesized fluctuations which generate local ice II-like, ice-III-like structures, etc. is not ridiculous. Perhaps the unusually large compressibility of water results from the existence of such structural fluctuations [86].

It is interesting to note that others have speculated on the possible role that multiple minima in the many molecule potential energy surface can play in determining the properties of liquid water [87]. Most of these models have, however, employed full, rigid, ice polymorph geometries rather than emphasizing the role of average geometries and the effects of coupled OOO angle and OO separation distributions. Nevertheless, the intellectual connection between the earlier speculations and ours must be acknowledged. Of particular interest, in view of the kind of random network we have proposed as a model of liquid H_2O, is the calculation by Kamb [75] of the radial distribution function for a mixture consisting of 50% ice Ih, 33% ice II and 17% ice III, each isotropically expanded 2% from the 100 K lattice dimensions, and with thermal broadening. The result is shown in Fig. 55. Clearly, testing of the random network model we have proposed will

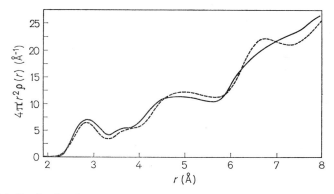

Fig. 55. Radial distribution function for liquid water synthesized as a mixture of ice I, ice II and ice III (from Ref. [75])

have to go beyond comparison of theoretical and observed pair correlation functions, although good agreement there is a necessary test every model must pass.

At least for the present the OH stretching region contour cannot be easily used to provide more than the qualitative information about variety of environments we have already discussed. In an early paper, Hornig and coworkers [88] used an empirical relationship between hydrogen bond length and frequency shift to estimate, from the observed decoupled OH band contour, the distribution of hydrogen bond lengths. More recently, Ford and Falk [89] have suggested a correlation between the distribution of local environments in liquid water, measured in terms of a distribution of potential energy, and the band contours of the stretching frequencies ν_1 and ν_3. They work out the quantitative details of such a correlation, computing the potential energy of liquid water and ice Ih from a combination of thermodynamic and spectroscopic data, and thereby infer that the infrared absorption contours of liquid HDO/H_2O and HDO/H_2O ice are consistent with a very narrow distribution of potential energy in the latter case, but a very broad distribution in the former case. However, variations in potential energy result from both hydrogen bond stretching and hydrogen bond bending, with the relative contributions of these unknown. In the liquid phase the overlapping effects of bond stretching and bond bending renders inversion of band contour data suspect except as a qualitative guide to thinking.

Theoretical studies, based entirely on the local tetrahedral pentamer model of the liquid [47] but including isotropic expansion and contraction with the distribution of OO separations matched to the observed OO correlation function, cannot completely account for the observed OH stretching spectrum [90]. The addition of some weak hydrogen bonds improves the predicted spectrum [90], but still leaves the widths of the band contours largely unaccounted for. It seems likely that inclusion of OOO angle distribution effects (*i.e.* bent hydrogen bonds as well as small dispersion about 109.5°) will improve the agreement between predicted observed spectra.

It will not have escaped the reader's notice that Walrafen's analysis [39] of the OH stretching region of the vibrational spectrum of the liquid also uses a

Gaussian decomposition of the contour into subbands, but that his assignments differ somewhat from ours. Walrafen also interprets the temperature dependence of the four dominant components of the contour in terms of the equilibrium between hydrogen-bonded and nonhydrogen bonded groups. We believe that interpretation to be inappropriate for $H_2O(as)$. Given the overall similarity between the spectra of liquid H_2O and $H_2O(as)$, one can question whether it is the best interpretation of the spectrum of the liquid. In fact, with some changes in terminology, the Walrafen analysis can be made consistent with the model we have proposed. Suppose the components of the OH stretching region of the liquid spectrum refer, as in the case of the model of $H_2O(as)$, to strongly bonded, bent bonded and weakly bonded molecules. Lack of suitable comparison data makes it impossible to check this assignment, but let us accept it for the purposes of argument. Then, surely, one must expect that as temperature increases, and with it molecular mobility, the variety of local environments corresponding to higher energies (taking a random network based on ice I as reference) will also increase. The observed temperature dependences of the component bands (see Section IV) follow the expected pattern. However, because of thermal expansion effects, it is not easy to convert observed intensities to weighted contributions to the structure [91]. It should also be clear that a quasi thermodynamic description of the contributions of the several local randomized environments is possible (see Section V.B), hence an "equilibrium between species" can be generated.

Our interpretation of the OH stretching region of the vibrational spectrum differs from Walrafen's primarily in the assertions that identifiable bent-bond environments are important and that intermolecular coupling is not likely to lead to a separate band. In the case of a dimer intermolecular coupling leads to a splitting of the symmetric and antisymmetric levels, hence usually (if selection rules permit) a splitting in the vibrational spectrum. This splitting is a reflection of precise phase relations between the two component zero order (independent molecule) wave functions. It seems unlikely to us that the superposition of interactions corresponding to the variety of environments in the liquid will preserve the phase relations sufficiently to permit resolution of the numerous splittings into only two bands. Of course, the coupling does lead to a spreading of the permissable vibrational excitations, but we think it likely this is subsumed into the width of the overall transition. If this argument is accepted, Walrafen's assignment of components must be changed; by using an argument like that in Section V.B this can be affected without prejudice to the remainder of his argument on the temperature dependence of the bands, etc. But why should the model we have proposed have a multimodal distribution of environments? Is it not possible (likely?) that molecular mobility and thermal excitation blur all environmental distinctions and that there is only a unimodal, broad distribution of environments? Perhaps, but our basic argument is that the many molecule potential energy surface is such as to generate a multi-modal distribution of environments. That this is so is implied by the existence of several crystalline forms of ice with the same nearest neighbor OO separation. We believe this feature of the surface to be very important, and suggest that it persists in the liquid to the extent that even though the distribution of environments is thermally broadened it cannot be adequately described as unimodal. It is because of our wish to exploit this property of the many molecule potential energy surface that we feel the introduction of "species", even when rendered equivalent to some other description as suggested above, obscures the fundamental structural properties of the liquid.

Finally, we must call attention to a conflict between the considerations of this section and one of the broad generalizations derived from model theories. We remarked in Section V that the lattice and cell models (and to some extent the cluster model) predicted very little change in local environment as the temperature changes. In contrast, our rewording of Walrafen's interpretation requires there to be an alteration in local environment as the temperature is changed. It is not hard to see where the error in the models arises. All the crude models (but not simulated water) oversimplify the many molecule potential energy surface by restricting attention to a small subset of the possible minima. But our interpretation of the liquid asserts that there is an important contribution to the structure arising from fluctuations which generate changes in the potential energy surface and in the distribution of positions and types of minima. To the extent this is valid, the simpler models are deficient.

We close this section by emphasizing again that the difference between the model proposed here, and earlier similar models, is the current emphasis on:

(i) the existence of a random network structure with parentage in several lattice types,

(ii) the importance of deviations from regularity in the several types of local environments,

(iii) the role of dynamics, *i.e.* the continuous fluctuations which generate the local environments and deny the possibility of stationary configurations of molecules.

Note added in proof. The notion that there are local environments in liquid H_2O which are distorted variants of those in ice II, ice III, etc., is consistent with some recent observations by H. Kanno, R. J. Speedy and C. A. Angell [Science *189*, 880 (1975)]. They show that the limiting temperature for homogeneous nucleation as a function of pressure parallels the equilibrium melting line even to exhibiting a break at the same pressure as does the equilibrium line, which break is defined by the intersection of the melting lines for liquid water $\leftrightarrow$ ice I and liquid water $\leftrightarrow$ ice III. Angell *et al.* interpret this behavior as an indication that compression of the liquid is effected by bending of hydrogen bonds, and that there is a tendency for the compression to create local environments with at least some resemblance to those in ice III.

Mention should also be made of the study of sound propagation in water, using the method of molecular dynamics, by Rahman and Stillinger [Phys. Rev. *10A*, 368 (1974)]. They find, in addition to ordinary sound wave propagation, the existence of an extra higher frequency sound wave which propagates with about twice the velocity of the ordinary sound wave. This ratio of sound velocities is very close to the ratio of longitudinal to transverse sound velocities in polycrystalline ice. The point is *not* that the comparison cited requires the existence of rigid ice-like structure in the liquid, which it does not. Rather, the importance of the observation is that it implies the existence of local correlated motion on the scale of a few molecular diameters, in turn implying that on this scale there must be some structure and not just randomly placed molecular centers and orientations.

VII. Some Tests

The discussion to this point can be summed up as follows. The proferred interpretations of the Raman spectroscopic data and of the X-ray and neutron diffraction data agree (a) with respect to the existence of local tetrahedrality in H_2O(as), implying near linearity of the hydrogen bonding between nearest neighbor water molecules, (b) with respect to the existence of a distribution of OOO angles with full width at half maximum estimated to be in the range in $\sim 6°$—$8°$ and centered at or near the tetrahedral value $109.5°$, and (c) with respect to the probable existence in high density H_2O(as) of some local environments characterized by very small OOO angles. Drawing on these conclusions, and noting that the existence of three crystalline phase (ices I, II, III) with the same nearest neighbor OO separation and with only slightly different free energies argues that small coordinated rotations and displacements of water molecules lead to potential surfaces with minima representing different local symmetries, it has been suggested that there is likely a remnant of this property in the potential surface for the liquid. (Recall that the hydrogen bond interaction is of the order of 10 k_BT at room temperature). That is, although the distribution of angular orientations is likely continuous, it need not be flat and featureless: there could be broad maxima centered around the orientations known to be local minima by virtue of the existence of the several ice structures. Clearly, the randomized structure with ice I — like symmetry has the lowest free energy at 1 atm, hence one would expect preferential population of this structure as the temperature is lowered. On the other hand as the temperature is raised, whatever broadened minima exist must be broadened still futher, and the populations of the several minima must tend to equalize. In this view high density amorphous solid water is a metastable version of "ordinary" liquid water. Presumably this is a consequence of the very low mobility of the water molecules under the conditions leading to the high density deposit. If correct, there may exist a continuum of higher density H_2O(as) deposits, corresponding to slightly different conditions of formation (molecular mobility), each one related to a slightly different thermodynamic state of the ordinary liquid, while the low density form of H_2O(as) is likely unique.

The preceding is equivalent to the speculation that liquid H_2O has a random network structure, specifically including some configurations with small OOO angle, but the large molecular mobility and the extensive thermal excitation lead to distributions of OOO angles and OO separations with much larger dispersions than their H_2O(as) counterparts. The consequences attendent to a large mobility are twofold. First, many local geometries are probed, hence many possible forms of connected (including badly distorted hydrogen bond network can be sampled. Second, the dynamics is such that any specified geometry is populated and depopulated rapidly. The net effect is the generation, instantaneously, of a continuous random network in which, we assert, there are same preferred small OOO angles, but (as already stated) with much larger dispersion of OOO angles and OO separations than is characteristic of the immobile random network of an amorphous solid. This model is conceptually different from those in the class of mixture models, but is not irreconcilable with them.

If the different continuous random network models of high and low temperature H_2O(as) are valid, the following tests are worthy of attention:

(i) Corresponding to the proposed different parentages of the continuous random networks laid down at ~10 K and ~77 K, there should be differences in the vibrational spectra. In particular, if the band at 3370 cm^{-1} or at 3220 cm^{-1} H_2O(as) is really associated with molecules with bent hydrogen bonds in local randomized ice II-like and/or ice III-like environments, it should not appear in the high temperature form of H_2O(as). On the other hand, if it is really associated with the overtone 2 ν_2 (or some other non bent bond source) it should appear in both forms of H_2O(as).

(ii) The high temperature form of H_2O(as) should have a negative coefficient of thermal expansion, the low temperature form a positive coefficient of thermal expansion.

(iii) At very low temperature the specific heat of high temperature H_2O(as) should exceed that of ice Ih, corresponding to the existence of structural configurations of nearly equal energy [92,93,94]. The existence of a transition to ice Ic at ~135 K suggest the barrier between configurations to be of the order of magnitude ~500 cal/mole. A similar statement could be made with respect to the low temperature specific heat of low temperature H_2O(as). However, in view of our structural hypothesis, it is not clear what crystalline form should be chosen for comparison.

(iv) The low and high temperature forms of H_2O(as) should have different dielectric constants to the extent that a continuous random network with (some) ice II parentage retains proton ordering at low temperature [11].

(v) Available experimental data hint at some sort of structural transformation in supercooled water near 228 K [81]. At this temperature the entropy and density of the liquid are barely more than those of ice Ih. Since the liquid has almost the same volume as the crystal it is reasonable to conjecture that it has a complete random network structure of unique parentage of ice I type. The librational motion of water molecules is of sufficiently high frequency that at 228 K the difference in the contributions to the thermodynamic properties of the liquid and the solid must be small. Perhaps the principle effect of the structural transformation is to freeze out the restricted translational motion. But the hindered translations in the crystal do not differ much in frequency from those in the liquid, hence the entropy change will be small.

(vi) If (v) is accepted, low temperature H_2O(as) is metastable with respect to high temperature H_2O(as).

(vii) Accepting (v), the vibrational spectrum of supercooled water should approach that of high temperature H_2O(as), and not that of low temperature H_2O(as).

Note added in proof. (viii) Suppose liquid water is excited by a short intense pulse of frequency selected infra-red radiation. Let the frequency be chosen to coincide with OH stretching in one of the inferred subcomponents (linear hydrogen bonds, bent hydrogen bonds, etc.). Finally, suppose the incident pulse is intense enough to appreciably depopulate the ground state (of that subcomponent). After a long time, one expects the vibrational population to be depleted by a bimolecular near resonant scattering of a quantum of OH stretching (ν_1) into two quanta of HOH bending (ν_2) as there is near resonance ($\nu_1 \approx 2\,\nu_2$). If this is the case there should be very rapid decay of the population of this subcomponent since the anharmonicity of bending HOH is small. In ethanol this process of energy transfer has a lifetime of $\sim$28 psec. [A. Laberau, G. Kehl and W. Kaiser, Opt. Commun. *11*, 74 (1974)]. However, of more importance, it is likely that before this time has elapsed the environment around the vibrationally excited molecule will relax. It is now believed [C. J. Montrose, J. A. Bucero, J. Marshall-Coakley and T. A. Litovitz, J. Chem. Phys. *60*, 5025 (1974)] that rotational relaxation occurs within 5 psec at 273 °K, and that hydrogen bond formation and breaking by translational motion is even faster, with a relaxation time of $\sim$0.6 psec at 273 °K. The use of super-cooled water and, presumably, also amorphous solid water, permits generation of conditions wherein these relaxation times are increased. A monitor pulse, incident a variable interval following the excitation pulse, can be used to probe the environments populated by relaxation from the initially prepared state. Now, if randomized bent hydrogen bond configurations approximately describe some of the complex environments in the liquid, one should see an environmental relaxation rate like that for rotational relaxation, sensibly the same for each subcomponent of the spectrum (neglecting fast relaxation of any Fermi resonance peak 2 ν_2). If, on the other hand, there are important contributions from broken hydrogen bond configurations which do not require rotational reorientation, but only a center of mass motion to regenerate the hydrogen bond, the relaxation rate should be much larger, indeed not observable with currently available picosecond pulse excitation technology. Furthermore, in the latter case relaxation will produce only environments with differences in hydrogen bond numbers. Clearly, intermediate cases are possible. In principle, then, by tuning picosecond infrared excitation pulsed across the vibrational absorption band and by lowering the water temperature some information can be obtained about the environments contributing to the broadening of vibrational transitions, hence some information about the structure which can be used to test the hypotheses discussed earlier. This experiment has been undertaken by M. G. Sceats and S. A. Rice.

Clearly, any measurement that differentiates between the properties of high and low temperature forms of H_2O(as), and/or delineates the relationship between H_2O(as) and liquid H_2O, can be used to test the hypotheses advanced *vis a vis* their structures. These and the experimental tests suggested, together with the construction of continuous random network models more sophisticated than that for Ge(as), the increased use of computer simulation, and exploitation of the available experimental information to guide the choice of apppproximations in a statistical mechanical theory should increase our understanding of H_2O(as) and, uitimately, liquid H_2O.

Conjectures on the Structure of Amorphous Solid and Liquid Water

Acknowledgements. This account could not have been written but for the contributions of Drs. D. Olander, C. G. Venkatesh, J. Wenzel, A. H. Narten and J. B. Bates. I wish to thank Dr. Mark Sceats and Mr. T. C. Sivakumar for stimulating questions and discussions. The experimental work on $H_2O(as)$ was carried out at the University of Chicago, the Danish Atomic Energy Authority Laboratory at Riso, and the Oak Ridge National Laboratory. In addition to provision of facilities by the laboratories mentioned, this research has been financed by a grant from the National Science Foundation (NSF GP 42930 X) and has derived benefit from the general support of materials research at the University of Chicago under NSF grant GH 33636A2MRL.

VIII. References

[1] Three excellent summaries of the experimental information available, and their theoretical interpretation are:
Eisenberg, D., Kauzmann, W.: The structure and properties of water. Oxford: University Press 1969.
Franks, F. (ed.): Water, a comprehensive treatise. New York: Plenum Press 1972.
Ben-Naim, A.: Water and aqueous solutions. New York: Plenum Press 1974.

[2] See, for example: Rice, S. A., Gray, P.: The statistical mechanics of simple liquids. New York: John Wiley & Sons 1965.
Croxton, C. A.: Liquid state physics. Cambridge: University Press 1974.

[3] Stillinger, F. H.: Advan. Chem. Phys., in press.
Rahman, A., Stillinger, F. H.: J. Chem. Phys. 55, 3336 (1971).
Stillinger, F. H., Rahman, A.: J. Chem. Phys. 57, 1281 (1972).

[4] Olander, D., Rice, S. A.: Proc. Natl. Acad. Sci. U.S. 69, 98 (1972).

[5] Burton, E. F., Oliver, W. F.: Proc. Roy. Soc. (London) A 153, 166 (1935).

[6] Wenzel, J., Linderstrom-Lang, C., Rice, S. A.: Science 187, 428 (1975).

[7] Venkatesh, C. G., Rice, S. A., Narten, A. H.: Science 186, 927 (1974).

[8] Venkatesh, C. G., Rice, S. A., Bates, J. B.: J. Chem. Phys. 63, 1065 (1975).

[9] Blackman, M., Lisgarten, N. D.: Advan. Phys. 7, 189 (1958) survey the early diffraction experiments. — Dowell, L. G., Rinfret, A. P.: Nature 188, 1144 (1960). — Blackman, M., Lisgarten, N. D.: Proc. Roy. Soc. (London) A 239, 93 (1957). — Beaumont, R. H., Chihara, H., Morrison, J. A.: J. Chem. Phys. 26, 1456 (1961).

[10] See also Venkatesh, C. G.: Ph. D. thesis, University of Chicago 1975.

[11] Fletcher, N. H.: The chemical physics of ice. Cambridge: University Press 1970.

[12] Steele, W. A., Pecora, R.: J. Chem. Phys. 42, 1863 (1965).

[13] Blum, L.: J. Computational Phys. 7, 592 (1971).

[14] Narten, A. H., Levy, H. A.: J. Chem. Phys. 55, 2263 (1971).

[15] For a more general discussion see Powles, J. G.: Advan. Phys. 22, 1 (1973).

[16] A very convenient summary is in Narten, A. H.: Oak Ridge National Laboratory Report ORNL 4578, UC-4-Chemistry (1970). See also Narten, A. H., Levy, H. A.: In: Water, a comprehensive treatise (F. Franks, ed.), Vol. I, p. 311. New York: Plenum Press 1972.

[17] Hendricks, R. W., Mardon, P. G., Shaffer, L. B.: J. Chem. Phys. 61, 319 (1974).

[18] Page, D. I., Powles, J. G.: Mol. Phys. 21, 901 (1971).

[19] Narten, A. H.: J. Chem. Phys. 56, 5681 (1972).

[20] Lengyel, S., Kalman, E.: Nature, 248, 405 (1974).

[21] Whalley, E.: Mol. Phys. 28, 1105 (1974).

[22] A general review of the properties of Ge(as) is found in Paul, W., Temkin, R., Connell, G. N.: Advan. Phys. 22, 529, 581, 643 (1973).

[23] Steinhardt, P., Alben, R., Weaire, D.: J. Noncrystalline Solids 15, 199 (1974).

[24] Duffy, M. G., Boudreaux, D. S., Polk, D. E.: J. Noncrystalline Solids 15, 435 (1974).

[25] Wenzel, J.: private communication.

[26] Bondot, P.: Compt. Rend. Acad. Sci. (Paris) B 265, 316 (1967).
Bondot, P.: Compt. Rend. Acad. Sci. (Paris) B 268, 933 (1969).

[27] Narten, A. H., Venkatesh, C. G., Rice, S. A.: submitted to J. Chem. Phys.

[28] Dean, P.: Proc. Phys. Soc. (London) 84, 727 (1964).

[29] Bell, R. J., Dean, P., Hibbins-Butler, D. C.: J. Phys. C 3, 2111 (1970).

[30] Bell, R. J.: J. Phys. C. 5, L 315 (1972).

[31] Hori, J.: Spectral properties of disordered chains and lattices. London: Pergamon Press 1968.

[32] Andersen, P. W.: Phys. Rev. 109, 1492 (1918).

[33] Economu, E. N., Cohen, M. H.: Phys. Rev. B 5, 2931 (1972).

[34] Bottger, H.: Phys. Stat. Solidi 62, 9 (1974).

[35] Weaire, D., Alben, R.: Phys. Rev. Letters 29, 1505 (1972).

[36] Alben, R., Smith, J. E., Brodsky, M. H., Weaire, D.: Phys. Rev. Letters 30, 1141 (1973).

[37] Herzberg, G.: Molecular structure and molecular spectra, Vol. II. Princeton, N. J.: D. Van Nostrand 1945.

[38] A convenient review of infrared data for water is Hornung, N. J., Choppin, G. R., Renovitch, G.: Appl. Spect. Rev. *8*, 149 (1974).

[39] Walrafen has published an extensive review of Raman studies of water: Walrafen, G.: In: Water, a comprehensive treatise (F. Franks, Ed.), Vol. 1, p. 151. New York: Plenum Press 1972.

[40] α is the polarizability and μ the dipole moment of the molecule.

[41] Cunningham, K., Lyons, P. J.: J. Chem. Phys. *59*, 2132 (1973).

[42] Scherer, J. R., Go, M. K., Kint, S.: J. Phys. Chem. *78*, 1304 (1974).

[43] See the discussion in Ref. [39].

[44] Colles, M. J., Walrafen, G. E., Hecht, K. W.: Chem. Phys. Letters *4*, 621 (1970).

[45] Senior, W. A., Verrall, R. E.: J. Phys. Chem. *73*, 4242 (1969).

[46] Falk, M., Wyss, H. R.: J. Chem. Phys. *51*, 5727 (1969).

[47] Curnutte, B., Bandekar, J.: J. Mol. Spectry. *41*, 500 (1972).

[48] Hardin, A. H., Harvey, K. B.: Spectrochim. Acta. *29A*, 1139 (1973).

[49] Buontempo, U.: Phys. Letters *42A*, 17 (1972).

[50] Hardin and Harvey (Ref. [48]) find $\sim$3191 cm^{-1} and $\sim$2372 cm^{-1} from infrared absorption.

[51] Li, P. C., Devlin, P.: J. Chem. Phys. *59*, 547 (1973).

[52] See, for example Ref. [42].

[53] Taylor, M. J., Whalley, E.: J. Chem. Phys. *40*, 1660 (1964).

[54] Hardin and Harvey find $\sim$3304 cm^{-1}.

[55] Rotational relaxation must depend to some extent on the molecular moment of inertia.

[56] See Ref. [11] and also Kamb, B.: In: Structural chemistry and molecular biology, p. 507 (eds. A. Rich and N. Davidson). San Francisco: W. H. Freeman 1968.

[57] Frank, H.: In: Water, a comprehensive treatise, Vol. 1, p. 515 (F. Franks, ed.). New York: Plenum Press 1972.

[58] Ben-Naim, A.: In: Water, a comprehensive treatise, Vol. 1, p. 413 (F. Franks, ed.). New York: Plenum Press 1972.

[59] Rowlinson, J. S.: Trans. Faraday Soc. *47*, 120 (1961).

[60] Ben-Naim, A., Stillinger, F. H.: In: Structure and transport processes in water and aqueous solutions (R. A. Horne, ed.). New York: Interscience 1972.

[61] Fleming III, P. D., Gibbs, J. H.: J. Stat. Phys. *10*, 157, 351 (1974).

[62] Bell, G. M.: J. Phys. C. *5*, 889 (1972).

[63] Weissmann, M., Blum, L.: Trans. Faraday. Soc. *64*, 2605 (1968).

[64] Weres, O., Rice, S. A.: J. Am. Chem. Soc. *94*, 8983 (1972).

[65] Pauling, L.: J. Am. Chem. Soc. *57*, 2680 (1935).

[66] Denley, D., Rice, S. A.: J. Am. Chem. Soc. *96*, 4369 (1974).

[67] Angell, C. A.: J. Phys. Chem. *75*, 3698 (1971).

[68] Hagler, A. T., Scheraga, H. A., Nemethy, G.: J. Phys. Chem. *76*, 3229 (1972).

[69] Lantz, B. R., Hagler, A. T., Scheraga, H. A.: J. Phys. Chem. *78*, 1531 (1974).

[70] Strictly speaking, the two models are not mutually compatible, and the concentrations of different cluster species are not equivalent to the weights associated with different basic lattice section occupancies. Nevertheless there exists the similarity that thermodynamic weights are associated, in each calculation, with various kinds of hydrogen bonded configurations of molecules. How these are subsequently used differs profoundly. I have deliberately used the very loose language of the text to extract for the reader the concept most nearly common to the two approaches.

[71] Sarkisov, G. N., Dashevsky, V. G., Malenkov, G. G.: Mol. Phys. *27*, 1249 (1974).

[72] Watts, R. O., Mol. Phys. *28*, 1069 (1974).

[73] Stillinger, F. H., Rahman, A.: J. Chem. Phys. in press.

[74] von Hippel, A., Farrell, E. F.: Tech. Rept. #9, Contract N00014-67-A-0204-00053, Lab. Insulation Research, MIT (1971).

[75] Kamb, B.: In: Structural chemistry and molecular biology, p. 507 (eds. A. Rich and N. Davidson). San Francisco: W. H. Freeman 1968.

[76] Stillinger, F. H., Lemberg, H.: submitted to J. Chem. Phys.

[77] Kell, G. S.: Can. J. Chem. *52*, 1945 (1974).

[78] Polk, D. E.: J. Noncrystalline Solids 5, 365 (1971).
Polk, D. E., Boudreaux, D. S.: Phys. Rev. Letters 31, 92 (1973).
Shevchik, N. J., Paul, W.: J. Noncrystalline Solids 8–10, 381 (1972).
Henderson, D., Herman, F.: J. Noncrystalline Solids 8–10, 359 (1972).
Steinhardt, P., Alben, R., Weaire, D.: J. Noncrystalline Solids 15, 199 (1974).
Duffy, M. G., Boudreaux, D. S., Polk, D. E.: J. Noncrystalline Solids 15, 435 (1974).

[79] Zachariasen, W. H.: J. Am. Chem. Soc. 54, 3841 (1932). — Ordway, F.: Science 143, 800 (1964). — Evans, D. L., King, S. R.: Nature 212, 1353 (1966). — Bell, R. J., Dean, P.: Nature 212, 1353 (1966).

[80] Pople, J. A.: Proc. Roy. Soc (London) A 205, 163 (1951).

[81] Angell, C. A., Shuppert, J., Tucker, J. C.: J. Phys. Chem. 77, 3092 (1973) and references cited therein. — Rasmussen, D. H., MacKenzie, P.: J. Chem. Phys. 59, 5003 (1973). — Zheleznyi, B. V.: Russ. J. Phys. Chem. 42, 950 (1968), 43, 1311 (1969).

[82] Alben, R., Boutron, P.: Science 187, 430 (1975).

[83] Narten, A. H.: private communication.

[84] Ghormley, J. A.: Science 171, 62 (1971).

[85] Sugasaki, M., Suga, H., Seki, S.: In: Physics of ice, p. 329. New York: Plenum Press 1969; Bull. Chem. Soc. Japan, 41, 2591 (1968).

[86] Frank, H. S.: In: Water, a comprehensive treatise, p. 519 (F. Frank, ed.). New York: Plenum Press 1972.

[87] For a brief review see Ref. [75].

[88] Wall, T. T., Hornig, D. F.: J. Chem. Phys. 43, (6), 2079 (1975).

[89] Ford, T. A., Falk, M.: Can. J. Chem. 46, 3579 (1968).

[90] Curnutte, B., Bandekar, J.: J. Mol. Spectry. 49, 314 (1974).

[91] This point should not be overemphasized. Useful information can be obtained from the temperature dependence of band intensities, as demonstrated by Walrafen.

[92] Zeller, R. C., Pohl, R. O.: Phys. Rev. B 4, 2029 (1971).

[93] Anderson, P. W., Halperin, B. I., Varma, C. M.: Phil. Mag. 25, 1 (1972).

[94] Phillips, W. A.: J. Low Temp. Phys. 7, 351 (1972).

Received January 15, 1975, October 6, 1975

Author Index Volumes 26—61

The volume numbers are printed in italics

Topics in Current Chemistry
Fortschritte
der chemischen Forschung

Editorial Board: A. Davison, M.J.S. Dewar,
K. Hafner, E. Heilbronner, U. Hofmann,
J.M. Lehn, K. Niedenzu, K. Schäfer, G. Wittig
Managing Editor: F. Boschke

11. Band, 1. Heft: **Lösungen und Lösungsmittel**
31 Abbildungen, 176 Seiten. 1968

Inhaltsübersicht: C. Reichardt, K. Dimroth:
Lösungsmittel und empirische Parameter zur
Charakterisierung ihrer Polarität. O. Fuchs:
Lösungen von makromolekularen Stoffen.
J. Falbe, B. Cornils: Oxo-Alkohole als Lösungs-
mittel. L. Rohrschneider: Der Lösungsmittel-
einfluß auf die gas-chromatographische Retention
gelöster Stoffe.

Volume 49: C.A. Mead: **Symmetry and Chirality**
23 figures. IV, 88 pages. 1974

For a systematic group-theoretical study of
chirality functions the chemist requires a knowledge of the representation theory of symmetric
and hyperoctahedral groups, and of the concept
of induced representations and their properties.
While excellent expositions of these topics are to
be found in the mathematical literature, they are
usually formulated in an idiom foreign to the
chemist and are thus relatively inaccessible to
him. As the present article is an attempt to bridge
this mathematical gap, the theory is presented
as far as possible in a unified form so as to include
the cases of both achiral and chiral ligands.
Readers are assumed to have only an average
chemist's knowledge of group theory.

Springer-Verlag
Berlin
Heidelberg
New York

Advances in Polymer Science

Edited by H.J. Cantow,
G. Dall'Asta, J.D. Ferry,
H. Fujita, M. Gordon,
W. Kern, G. Natta,
S. Okamura,
C.G. Overberger, W. Prins,
G.V. Schulz, W.P. Slichter,
A.J. Staverman, J.K. Stille

Advances in Polymer Science
comprises reports of mono-
graph type, dealing with
progress achieved in the
physics and chemistry of
high polymers, and includes
full bibliographical data. Its
objectives are to provide
those working in this field
with information on subjects
that are of special topical
interest and to report recent
advances that have been so
rapid as to require review-
type treatment.

Vol. 12:
62 figures. III, 190 pages
1973. Cloth

Contents: K. Osaki:
Viscoelastic Properties of
Dilute Polymer Solutions.
W.L. Carrick: The Mecha-
nism of Olefin Polymeri-
zation by Ziegler-Natta
Catalysts. C. Tosi,
F. Ciampelli: Applications
of Infrared Spectroscopy
to Ethylene-Propylene
Copolymers. K. Tsuji:
ESR Study of Photodegra-
dation of Polymers.

Vol. 13: W. Wrasidlo
Thermal Analysis
of Polymers
49 figures. II, 99 pages. 1974
Cloth

Contents: Scope of the
Review. Experimental
Methods. Glass Transitions.
Melting. Crystallization.

Vol. 14:
11 figures. III, 130 pages
1974. Cloth

Contents: J.P. Kennedy,
S. Rengachary: Correlation
between Cationic Model and
Polymerization Reactions of
Olefins.
J. Hutchison, A. Ledwith:
Photoinitiation of Vinyl
Polymerization by Aromatic
Carbonyl Compounds.
L.S. Gal'braikh,
Z.A. Rogovin: Chemical
Transformations of Cellulose

Vol. 15:
32 figures. III, 155 pages
1974. Cloth

Contents: G. Henrici-Olivé,
S. Olivé: Oligomerization
of Ethylene with Soluble
Transition-Metal Catalysts.
A. Zambelli, C. Tosi:
Stereochemistry of Propy-
lene Polymerization.
C.-D.S. Lee, W.H. Daly:
Mercaptan-Containing
Polymers. Yu. V. Kissin:
Structures of Copolymers
of High Olefins.

Vol. 16:
W.W. Graessley
The Entanglement
Concept in
Polymer Rheology

36 figures. II, 179 pages
1974. Cloth

Contents: Chain Configu-
ration in Amorphous Polymer
Systems. Material Properties
of Viscoelastic Liquids.
Molecular Models in Polymer
Rheology. Experimental
Results on Linear Viscoelastic
Behavior. Molecular Entan-
glement Theories of Linear
Viscoelastic Behavior.
Entanglement in Cross-linked
Systems. Non-linear Visco-
elastic-Properties.

**Springer-Verlag
Berlin
Heidelberg
New York**